Mössbauer spectroscopy and its applications

Mössbauer spectroscopy and its applications

T.E. CRANSHAW, B.W. DALE, G.O. LONGWORTH

UK Atomic Energy Research Establishment, Harwell

and

C.E. JOHNSON

Professor of Physics, University of Liverpool

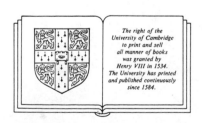

The right of the
University of Cambridge
to print and sell
all manner of books
was granted by
Henry VIII in 1534.
The University has printed
and published continuously
since 1584.

CAMBRIDGE UNIVERSITY PRESS
Cambridge
London New York New Rochelle
Melbourne Sydney

Published by the Press Syndicate of the University of Cambridge

The Pitt Building, Trumpington Street, Cambridge CB2 1RP

32 East 57th Street, New York, NY 10022, USA

10 Stamford Road, Oakleigh, Melbourne 3166, Australia

First published 1985

Printed in Great Britain at the University Press, Cambridge

British Library Cataloguing in Publication Data
Mössbauer spectroscopy and its applications.
1.Mössbauer spectroscopy
I.Cranshaw,T.E.
537.5'352   QC491

ISBN 0 521 30482 2
ISBN 0 521 31521 2 pbk

Library of Congress Cataloguing in Publication Data
Main entry under title

Mössbauer spectroscopy and its applications.

   1. Mössbauer spectroscopy. I. Cranshaw, T.E.
QD96.M6M637 1985   537.5'352   85_17511

ISBN 0 521 30482 2
ISBN 0 521 31521 2 (pbk)

CONTENTS

FOREWORD

It is more than 25 years since R.L. Mössbauer carried out the pioneering experiments on Resonance Spectroscopy of Gamma-rays which developed into the now well established powerful and versatile tool, known as Mössbauer Spectroscopy, for the study of solid state properties. The intrinsically high energy sensitivity of the technique and the relatively simple instrumentation needed, have led to the general availability of the technique and to a wide range of applications in many different fields.

The present volume aims at providing an introduction to the fundamental phenomena of the Mössbauer effect, to the experimental methods, and to the applications of Mössbauer Spectroscopy in many fields, including Solid State Physics and Chemistry, Metallurgy, Magnetism, Radiation Damage, Surface Science, Biochemistry, and Archaeology. The book should serve as an introductory text to final year undergraduates and to graduate students, and should provide the specialist with information outside his immediate field of interest.

P.B. Hirsch

Oxford 1985

CHAPTER I

THE MÖSSBAUER EFFECT

Resonance Absorption of Radiation by Free Atoms

The resonance absorption of radiation is a phenomenon well
known in many branches of physics. The excitation of a tuning
fork by sound at its resonance frequency, the scattering of
sodium light by sodium vapour and the excitation of a dipole
by radiofrequency radiation are some familiar examples. It
might be thought that the same phenomenon should occur for the
γ-radiation emitted when nuclei in excited states lose their
energy by radiation. However, the effect of the recoil momentum, which can be neglected for sound and light, becomes
dominant for γ-radiation, because of its much higher energy.
This may be illustrated by comparing typical cases of γ-emission
and optical emission.

Consider the emission of radiation from free atoms. Let the
energy of the excited state be E_1 above that of the ground
state, and the energy of the photon be E_γ. Then the momentum
of the photon is $p = E_\gamma/c$. By the conservation of momentum
this is the momentum p_R of the recoiling atom, so the recoil
energy E_R of the atom is

$$E_R = \frac{p_R^2}{2M} = \frac{E_\gamma^2}{2Mc^2} \approx \frac{E_1^2}{2Mc^2} \qquad (1.1)$$

where M is the mass of the atom. Thus

$$E_\gamma = E_1 - E_R \approx E_1 - \frac{E_1^2}{2Mc^2} \qquad (1.2)$$

In absorption, the recoil momentum is in the opposite direction,
so that the absorbed energy is less than E_1 by twice the recoil
energy i.e. E_1^2/Mc^2.

In all this it has been assumed that the emitting atom is at rest. In a gas the motion of the atoms will broaden the energy of the γ-rays. Let us suppose that an atom has velocity v making an angle θ with the direction of emission. Then the energy of the photon is changed by the Doppler effect to E'_γ so that

$$E'_\gamma = E_\gamma + E_\gamma \frac{v}{c} \cos \theta \qquad (1.3)$$

The expressions become simpler if we write $\epsilon = \frac{1}{2}Mv^2$ for the kinetic energy of the atom. Then

$$\Delta E_\gamma = E'_\gamma - E_\gamma = E_\gamma \frac{v}{c} \cos \theta = 2\sqrt{\epsilon E_R} \cos \theta \qquad (1.4)$$

Then if we allow θ to take all possible values, the spectrum is broadened by an amount

$$\Delta = 2\sqrt{\bar{\epsilon} E_R} \qquad (1.5)$$

Where $\bar{\epsilon}$ is the mean value of ϵ. Table I shows the values of E_1 E_R and Δ for typical cases of optical emission (E = 2eV, equivalent to a wavelength of 6000Å) and γ-ray emission (E = 100 keV), assuming as mass M = 100 and $\bar{\epsilon} \simeq 0.025$ eV corresponding to room temperature.

TABLE I

	optical	γ-ray
	eV	eV
E_1	2	10^5
E_R	2×10^{-11}	10^{-1}
Δ	10^{-6}	10^{-1}

We see that in the optical case, the recoil energy is negligible compared with the Doppler broadening, whereas in the case of γ-emission, the two are comparable. The comparison is more striking when we note that the "natural width" of the lines may be closely the same.

The "natural width" can be envisaged in the following way. A decaying nucleus produces an electric field at a point x given by

$$\xi_o(x,t) = \xi_o \exp \{i(\omega_o t - \kappa x) - \Gamma t/2\} \tag{1.6}$$

where ω_o and κ are the angular frequency and wave vector of the radiation. Then the probability P(t) of detection of the decay is given by $|\xi(t)|^2$ which decreases with a lifetime τ = 1/Γ. Quantum mechanically the natural width Γ of the state obeys a relation similar to the Heisenberg Uncertainty relation,

$$\Gamma \tau \geq \hbar \tag{1.7}$$

where τ is the lifetime for decay of the excited state. In both optical and γ-ray cases, we might have a lifetime of 10^{-7} ; corresponding to $\Gamma \sim 10^{-8}$ eV. Then for the optical case the recoil energy is negligible. The Doppler broadening is somewhat greater than the line width, and may be observed in suitably designed experiments. For the γ-ray case, on the contrary, both the Doppler broadening Δ and the recoil energy E_R are much larger than the line width Γ and resonance absorption may not be observable.

Before the discovery of the Mössbauer effect, some successful experiments on nuclear resonance absorption had been carried out. These experiments made use of two ideas: (i) the energy difference $2E_R$ may be directly supplied by utilising the Doppler effect as shown in Figs. 1. 1a and 1b. The velocities required are high ($\sim 10^2 - 10^3$ m s^{-1}), and were obtained by two methods: The emitting nuclei may be mounted on a high speed rotor, or the nuclei may be given their velocity by recoiling from a prior nuclear disintegration. (ii) the thermal broadening may be increased by heating the source and absorber, so that some overlap of the lines occurs, as shown in Fig. 1.2. In 1957, R.L. Mössbauer (1958) carrying out experiments of this kind on the nucleus [191]Ir found that lowering the temperature increases the absorption rather than decreasing it. This was due to a previously unsuspected effect which Mössbauer was able

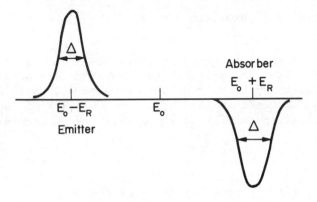

Fig. 1.1a Recoil energy prevents the
observation of NRA (Nuclear
resonance absorption)

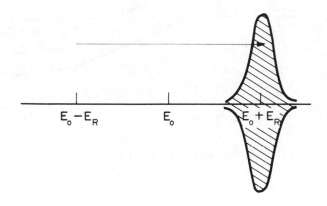

Fig. 1.1b NRA may be observed if the γ-ray
energy is Doppler shifted by $2E_R$

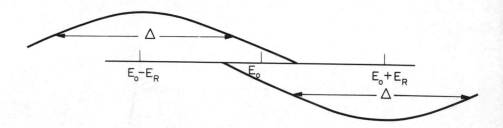

Fig. 1.2 NRA may be observed if the source
and absorber are heated

to explain, and which is now known as the Mössbauer Effect.

Resonance Absorption of Radiation by Atoms Bound in a Solid

(i) Classical description

The Mössbauer effect arises when we consider emitting atoms which are bound in a crystal and we take into account the fact that they are no longer able to recoil individually. A full account of the energy distribution of the radiation requires a quantum mechanical treatment, but we first give a classical account which brings out several of the essential points.

A classical radiator does not experience the recoil effect, but the distribution in frequency is entirely governed by the Doppler effect. Thus we return to (1.6) and neglecting the effect of decay

$$\xi = \xi_0 \exp\{i(\omega_0 t - \kappa x)\} \tag{1.8}$$

gives the disturbance due to an emitter at x. If x = vt, we have

$$\xi = \xi_0 \exp\{i(\omega_0 - \kappa v)t\} \tag{1.9}$$

which is the expression for the Doppler effect. Now suppose v = a cos ωt, representing a source oscillating harmonically about the origin with velocity amplitude a and frequency ω. Then

$$\xi = \xi_0 \exp(i\,\omega_0 t)\exp(-i\kappa a \cos \omega t)t$$

$$= \xi_0 \exp\{i(\omega_0 - a\kappa \cos \omega t)t\} \tag{1.10}$$

This is the expression for a frequency modulated wave with the "modulation index" $m_\omega = \frac{a\kappa}{\omega}$, which can be Fourier analysed into a component with frequency ω_0 and "sidebands" at frequencies $\omega_0 \pm n\omega$ where n = 1,2 ...... If m_ω is small, the standard theory shows that the intensity of radiation with frequency ω_0 is $J_0^2(m_\omega)$, or f, the fraction of the radiated energy with frequency ω_o, is given by

$$f = J_o^2 (\kappa x_o) \tag{1.11}$$

where x_o is the peak displacement of the atom from its equilibrium position and J_o is the Bessel function of order one. This is the answer for an Einstein crystal, where atoms are represented by harmonic oscillators of frequency ω_E.

In a more realistic model, there will be a large number of frequencies, and the expression for the intensity in the undisturbed line becomes

$$f = \prod_m J_o^2 (\kappa x_m) \tag{1.12}$$

where x_m is the peak amplitude for the mth mode. Since x_m is small for each value of m, we can make use of the expansion

$$J_o(y) = 1 - \frac{1}{4} y^2 \tag{1.13}$$

and then

$$\ln f = 2 \sum_m \ln J_o \sim 2 \sum_m \ln \{1 - \frac{1}{4} (\kappa x_m)^2\}$$

$$\sim = 2 \sum_m \frac{1}{4} \kappa^2 x_m^2 \tag{1.14}$$

For harmonic oscillation, the mean square deviation is given by $<x^2> = \frac{1}{2} \sum_m x_m^2$ so that

$$f = \exp (-\kappa^2 <x^2>) \tag{1.15}$$

(ii) Quantum description

In a quantum mechanical description, the physical origin of the Mössbauer Effect, i.e. of the possibility of momentum transfer without energy transfer, lies in the quantum nature of lattice

vibrations. On the Einstein model of a solid the lattice can
only have an energy 0, $\hbar\omega_E$, $2\hbar\omega_E$, ... and so can only change
its energy in units of $\hbar\omega_E$. So if E_R is less than $\hbar\omega$ the
lattice cannot absorb the recoil energy, and the γ-ray is
emitted with energy E_1. On the Debye model there is a spectrum
of frequencies ω up to a maximum cut-off frequency ω_D. The
conclusion when E_R is less than $\hbar\omega_D$ is the same as for the
Einstein frequency, but we now have to consider the possibility
of the excitation of the lattice (phonons) at a lower frequency,
say ω_D/N. It turns out that this involves the excitation of N
atoms together, so that the recoil energy to be absorbed is
now smaller and is E_R/N, which is again too small to excite the
lattice.

The detailed quantum mechanical calculation shows that the
probability of recoilless emission on absorption is

$$f = \exp\ (-\kappa^2\ \langle x^2 \rangle)$$

the same result as previously obtained by a classical treatment.

The recoilless fraction f

The probability of emission (or absorption) of γ-rays is given
by (1.15). Before proceeding to a calculation of $\langle x^2 \rangle$ for
crystals, it will be instructive to consider the implications
of (1.15) in some more unusual cases. The value of $\langle x^2 \rangle$ with
which we are concerned is the average over a time of the order
of the lifetime of the nuclear state, e.g. $\sim 10^{-7}$ s for ^{57}Fe.
For a typical low energy γ-ray $\kappa \sim 5 \times 10^{10}$ m^{-1}. The case for
an optical level with $\kappa \sim 10^7$ m^{-1} was discussed by Dicke (1952)
long before the discovery of the Mössbauer effect. He showed
that in a gas the restriction of $\langle x^2 \rangle$ by collisions with other
atoms can result in a strong narrowing of the line, and as a
simple model, Dicke considered an atom moving in a box. This
situation is approximately realized for γ-rays by the radiation
emitted from krypton in a clathrate compound, where the Kr
atom is trapped inside a large molecule, and only weak van der
Waals type forces exist between it and its surrounding. $\langle x^2 \rangle$ is
thus fixed by the structure of the molecule. It is found that
over a considerable temperature region, f is independent of
temperature. (Hazony et al. 1962, Steyert and Craig, 1962).
We see that it is the mean square displacement of the atom
which determines f, rather than its velocity or energy. As the

temperature is lowered, and the thermal energy becomes small compared with the binding energy, then the value of f starts to increase because the atom, so to speak, "sticks to the wall". A similar effect is found with Fe dissolved in In.

To estimate the value of $<x^2>$, we need a model of the crystal lattice, and various levels of approximation are discussed in Chapter IV on lattice dynamics. For the Einstein model $<x^2> = \kappa T/m\omega^2$ at high temperatures so that

$$f = \exp\left(-\kappa^2 \frac{kT}{M\omega^2}\right)$$

where ω is the characteristic frequency. For the Debye model at high temperatures

$$f = \exp\left(-6 \frac{E_R T}{k\theta_D^2}\right)$$

where $\theta_D = h\omega_D/k$ is the characteristic Debye temperature.

With zero point motion included, the Debye model gives the probability f at any temperature T that a nucleus will emit (or absorb) a γ-ray without losing (or gaining) energy from the lattice as

$$f = \exp\left\{ - \frac{3}{2} \frac{E_R}{k\theta_D}\left[1 + 4 \left(\frac{T}{\theta_D}\right)^2 \int_o^{\theta_D/T} \frac{x\ dx}{e^x - 1}\right] \right\} \quad (1.16)$$

As the temperature T increases, the fraction f decreases. We can now see the explanation of Mössbauer's result on Ir[191]. The energy distribution for the emitted and absorbed γ-rays in a solid are as shown in Fig. 1.3. There is a zero-phonon peak whose intensity decreases with increasing temperature. Resonant absorption therefore takes place without any Doppler shift, but the effect decreases as the source and absorber are heated in contrast to the free atom case Fig. 1.2. Conversely as the temperature decreases f increases, but not indefinitely owing to zero-point energy which limits it to a maximum of $\exp\left(-\frac{3}{2} \frac{E_R}{k\theta_D}\right)$.

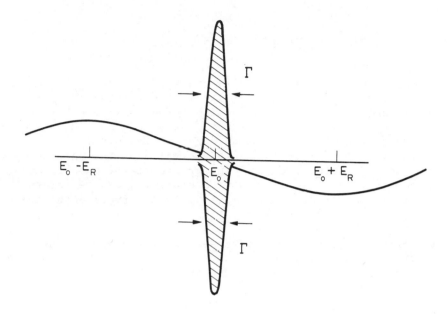

Fig. 1.3 For a bound nucleus when $E_R \ll k\theta_D$
there is a zero-phonon peak

The fundamental condition necessary for the Mössbauer Effect
to occur is that the root mean square displacement of the
nuclei (or atoms) in a solid should be less than the wavelength
of the γ-radiation (see (1.15)).

The essential practical condition necessary to observe the
Mössbauer Effect is $E_R < k\theta_D$ and we see thst we need γ-rays of
low energy and solids with a high Debye temperature, i.e. large
binding energy. Since Debye temperatures are of the order of a
few hundred degrees the γ-ray energy must be of the order of
10 - 100 keV. The Mössbauer Effect has now been observed in
about 100 nuclei of which the best known are ^{57}Fe (E_γ = 14.4
keV) and ^{119}Sn (E_γ = 26 keV).

Cross-section

The ease with which the Mössbauer Effect may be observed stems from the large value of the nuclear absorption cross-section σ_o at resonance. This is given by

$$\sigma_o = \frac{1}{1+\alpha} \; \frac{2I_e+1}{2I_g+1} \; \frac{2\pi}{\kappa^2} \qquad (1.16)$$

So that when κ^2 is small (as is required to make f large) σ_o will be large. Indeed σ_o for the 14.4 KeV γ-ray of ^{57}Fe is enormous ($\sim 10^6$ barns) and is much larger than the cross-section for any other absorption process e.g. by the photo-electric effect or the Compton effect. The statistical factor $(2I_e+1)/(2I_g+1)$ is just the ratio of the degeneracy of the excited state to that of the ground state, when I_e and I_g respectively are the spins of the states, α is the internal conversion coefficient i.e. $1/(1+\alpha)$ is the fraction of transitions which occur by photons.

In terms of the resonant cross-section the shape of the absorption line is given by the Breit-Wigner formula for the cross-section σ as a function of energy E

$$\sigma = \sigma_o \; \frac{\Gamma^2/4}{(E-E_\gamma)^2+\Gamma^2/4} \qquad (1.17)$$

Since the emission line has an identical shape, the observed line will be the convolution of the source and absorber lines which is a Lorentzian of width (full width at half maximum) 2Γ.

Apparatus for Mössbauer Spectroscopy

The Mössbauer Effect (recoilless nuclear transition) thus enables the resonant absorption of γ-rays of "natural width" to be observed. Nuclei in solids have their energy levels split or shifted by internal (hyperfine) interactions but these have previously been undetectable by γ spectroscopy because of the thermal broadening of the line. The discovery of the Mössbauer Effect makes possible the observation of these splittings, so that we now have a useful and valuable experimental tool for probing the electronic environment of the nuclei.

The conventional type of x-ray spectrometer uses a crystal lattice set at the Bragg angle to the x-ray beam to achieve dispersion, rather in the way that a grating may be used at optical wavelengths. The resolution obtainable in this way is far from sufficient to resolve the details of the Mössbauer spectrum. The hyperfine interactions which we wish to study are typically about 10^{-6} eV, whereas the energy of the γ-ray is of the order of 10^4 eV. Thus a resolution of 1 part in 10^{10} is required. Indeed, to resolve the line-shape requires a resolution of 1 part in 10^{12}. The only way to achieve this resolution is to make use of the resonance phenomenon. A Mössbauer spectrometer therefore works less like a dispersive spectrometer and more like a radio receiver in which a resonant circuit is tuned over the range of frequencies of interest and its response noted. The almost universal way of achieving the "tuning" is by use of the Doppler effect. It is easy to calculate the order of magnitude of velocity required, for $\delta\nu/\nu = v/c$. Thus in the example given above, to observe the hyperfine interactions requires a velocity of about 30 mm s^{-1}. In practice, spectra of ^{57}Fe compounds require velocities of the order of $\pm$ 8 mm s^{-1}, and spectra of most other nuclei require velocities less than 40 mm s^{-1}. A few cases are known where the velocity needed is as large as 600 mm s^{-1}.

Let us therefore imagine that we have a compound containing an element which is known to possess a Mössbauer nucleus, i.e. a nucleus with a suitable low lying level, and we wish to obtain the Mössbauer spectrum. The first step is to obtain a source of the radiation emitted when the nucleus falls from the excited state to the ground state. This problem is most conveniently solved if there exists a radioactive element which decays into the Mössbauer nucleus, populating the excited state. This is the case with ^{57}Fe, for there exists the species ^{57}Co which decays with a lifetime of 270 days by K-capture into the 136 keV level in ^{57}Fe. This in turn decays with a life time of 10^{-8} s partly to the ground state, and partly to the 14.4 keV level. This level decays partly by emitting a γ-ray and partly by emitting a K "conversion" electron. The scheme is shown in Fig. 1.4, and is typical of several other Mössbauer nuclei. Other methods of exciting the Mössbauer level exist, and may be employed either because no suitable parent nucleus exists, or for some special purpose. Among these we may mention Coulomb excitation, nuclear reaction, e.g. d-p in the nucleus with one less neutron than the Mössbauer nucleus, or neutron capture in such a nucleus.

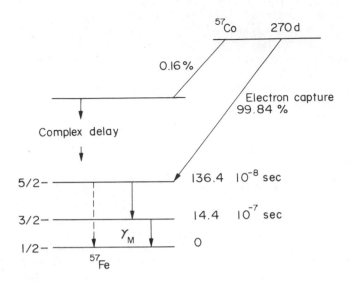

Fig. 1.4 The decay scheme of ^{57}Co

We must now find some way of imparting known velocities to the
source so that the energy can be shifted in a known way. A
simple direct method is to mount the source on a screw
mechanism similar to that on a screw cutting lathe. This
provides fixed velocities which can be selected. A more
convenient method is shown in Fig. 1.5 and uses a mechanism
which moves the source cyclically covering all the wanted
velocities in each cycle. An electronic circuit connected to
a small computer memory then routes the information to
different parts of the memory, each part corresponding to a
different velocity. The accumulating information can be
displayed on a cathode ray screen, and printed out at any time.
The waveform during the cycle may be sinusoidal. This has the
advantage of simplicity, but the disadvantage that most time
is spent at the extremes of velocity. A more convenient waveform is triangular or sawtooth, for then equal times are spent
at all velocity increments.

The most commonly employed mechanism for driving the source is
a moving coil vibrator. Usually some form of velocity
transducer is used to measure the velocity, and a servo-
mechanism employed to make the velocity follow the chosen

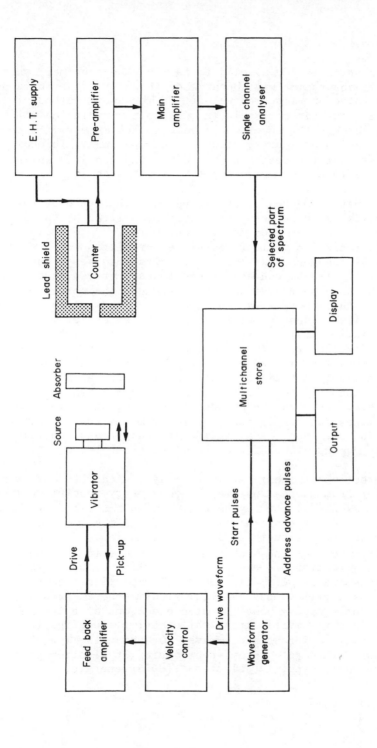

Fig. 1.5 A block diagram of a typical Mössbauer Spectrometer

waveform.

The radiation is now allowed to fall on the specimen, where resonant absorption will occur when the source has the appropriate value or values of velocity. The usual method of detecting the absorption is to measure the radiation which passes through the specimen. The computer memory will then contain an absorption spectrum of the specimen. This of course requires that the specimen should be prepared in a suitable thickness to allow the γ-rays to pass through it.

An alternative method of detecting the occurrence of resonance absorption is to measure the radiation emitted by the nuclei of the specimen as they decay back to the ground state. Part of this radiation will of course be γ-radiation of the same energy as the Mossbauer γ-ray which was absorbed. Usually, there will also be conversion electrons, x-rays, Auger electrons etc. When these radiations are detected, the spectrum accumulated in the memory gives the resonance frequencies and their intensities. These methods, called "back-scatter methods" have the advantage that the specimen may have any thickness. If conversion electrons (whose energy is given by the γ-ray energy less the binding energy of the electron in the atom) are detected, only nuclei very near the surface can be detected, because the range of the electrons is only a few hundred angstroms. This may be an advantage enabling surface layers to be examined and minimizing signals from the substrate. By selecting electrons in a particular energy band, nuclei at certain depths are preferentially detected.

The detection of the γ-rays, or x-rays is carried out by methods familiar from nuclear physics. Good energy resolution is not always essential, but may be needed when sources free from unwanted radiation are not available, and the specimen contains rather few of the nuclei under investigation. Then picking the signal out of the background may be difficult, and a careful selection of detector is essential. The conversion electrons are usually detected by a proportional counter using helium as a filling gas. Helium at atmospheric pressure has sufficient stopping power to stop electrons of a few keV energy in a few millimetres, while having a negligible sensitivity to x-rays and γ-rays. Occasionally the electrons are passed through a β-ray spectrometer of electromagnetic or electrostatic type, to select particular energies.

Other pieces of equipment in common use are cryostats for
obtaining low temperatures, down to 4.2K by using liquid helium
or 77K by using liquid nitrogen, and furnaces for high temperatures. Several experiments have also been made using He^3-
He^4 dilution refrigerators for temperatures in the milli-degree
range. Superconducting magnets providing fields of the order
of 10T are also commonly employed.

REFERENCES

Dicke, R.H. (1953). Phys. Rev. 89 472.
Hazony, Y., Hillman, P., Pasternak, M., Ruby, S. (1962). Phys.
  Lett. 2 237.
Mössbauer, R.L. (1958). Z. Physik 151 124.
Steyert, W.A. and Craig, P.P. (1962). Phys. Lett 2 165.

CHAPTER II

PROPERTIES OF RADIATION

The first kind of electromegnatic radiation to be deeply
studied was visible light. Newton showed that white light
could be separated into components, which we now consider
to be radiation of different wavelength, frequency or in
quantum mechanical language, energy. We do not wish here to
give a history of the theories of wave-particle duality. It
will be sufficient to observe that when light is detected, it
is detected by a quantum mechanical interaction, for example
photoemission, but when we ask how the energy is transported,
we have to adopt the standpoint of a wave theory. We are then
able to account for all the well known optical effects, and
the operation of lenses, prisms, gratings interferometers, etc.

With the advent of radio in the twentieth century, and the
production of electronic circuits capable of maintaining
oscillations at frequencies between 10^6 and 10^8 Hz, a different area of study of the electromagnetic radiation was
developed. Here, quantum mechanical processes play a very
minor role, or no role at all. Instead, we may be concerned
with the consequences of modulating the radiation, and the
relationship between changes in time and changes in the spectral
distribution, effects which are well represented by the methods
of Fourier transformation. As is well known, modulation at a
frequency ν causes the appearance of sidebands above and below
the carrier frequency, and separated from it by frequency
differences of ν. Such effects are not easily observable in
the case of visible light because, at least until the invention
of lasers, the feasible modulation frequencies are only about
1 part in 10^8 of the radiation frequency, and leave the
sidebands within the "natural width" of the line, or the
resolution of the spectrometer, or both.

In the x-ray and γ-ray region of the spectrum the frequencies
are above 10^{18} Hz, and present an entirely different situation.
Here the wavelengths are so short that atomic lattices are the

only possible diffraction structures, and then only for the
long wavelength end of the spectrum. At the higher frequencies,
the quantum energies are of the order of 10^6 eV or greater,
compared with about 1 eV for visible light, and the radiation
behaves much more like a particle than a wave. It makes
collisions with electrons which, although they have to be
treated relativistically, resemble billiard ball collisions.

A fascinating paradox is presented by the Mössbauer Effect.
We have available sufficient frequency resolution to reveal
the spectral changes produced by easily generated modulation
frequencies, just as we have at radio frequencies, but now in
the x-ray region where quantum effects predominate. This
leads to many conceptual difficulties which are perhaps not
entirely resolved at the present time. In this chapter, we
describe some striking experiments the results of which seem
to have entirely vindicated the wave-particle approach.

The birth of the 14 keV level in ^{57}Fe is signalled by the
emission of a 122 keV γ-ray. This provides us with a time t=0.
From this time onward, the excited state wave function and the
probability of detecting the 14 keV photon decay exponentially
to zero. Two types of experiment have been devised. In the
first, we arrange that the detector is sensitive between two
times T_1 and T_2 instead of from t=0 to t=∞. We then ask what
is the spectral distribution of the radiation so detected?
In the second, we pass the radiation through a resonant medium
which distorts the spectrum, and ask what is now the time
dependence of the probability of detecting a γ-ray. We treat
these two in turn.

Let us first consider the case when the detector sensitivity
is unrestricted. Then we may write the electric field
produced by the nucleus regarded as an oscillating dipole

$$\xi(x,t) = \xi_o \exp\{i(\omega_o t - \kappa x) - \Gamma t/2\} \tag{2.1}$$

This describes a wave with angular frequency ω_o, travelling in
the x direction with velocity $c = \omega_o/\kappa$, whose amplitude decays
exponentially.

Neglecting the spatial variable, the spectral distribution of
the radiation can be obtained by taking the Fourier transform

$$a(\omega) = \frac{1}{2\pi} \int_{0}^{\infty} \exp(i\,\omega_0 t - \Gamma t/2)\, \exp(-i\omega t)\, dt$$

$$= \frac{1}{2\pi} \frac{1}{(\omega - \omega_0) + i\Gamma/2} \qquad (2.2)$$

The intensity distribution is given by $I(\omega) \propto |a(\omega)|^2$

$$I(\omega) = \frac{1}{(\omega_0 - \omega)^2 + \Gamma^2/4} \qquad (2.3)$$

This is the well-known Lorentz distribution. It has a maximum for $\omega = \omega_0$, and falls to one half its value when

$$\omega_0 - \omega = \Gamma/2$$

Thus the full width of the distribution at half height is Γ. Now let us consider the case when photons are detected only up to a time T. Then the upper limit in the integral is T, and we find

$$I(\omega) = \frac{1 + e^{-\Gamma T} - 2e^{-\Gamma T/2} \cos(\omega - \omega_0)T}{(\omega - \omega_0)^2 + \Gamma^2/4} \qquad (2.4)$$

If T is short compared with $1/\Gamma$, this reduces to

$$I(\omega) = \frac{(1 - \cos(\omega - \omega_0)T)}{(\omega - \omega_0)^2}$$

$$= 4 \left\{ \frac{\sin(\omega - \omega_0)T/2}{(\omega - \omega_0)} \right\}^2 \qquad (2.5)$$

and the width at half height is found to be approximately 1/T.
The spread of the spectral distribution is much increased if
observations are restricted to times less than T, where
$T \ll 1/\Gamma$.

To try to measure this frequency distribution, we would employ
a Mössbauer spectrometer. This would mean that we would use
a resonant absorber whose resonance frequency could be changed
by utilizing the Doppler effect, and we would measure how much
radiation was transmitted as a function of Doppler velocity.
A complete description of such an experiment therefore involves
a calculation of the interaction of the modified radiation
with a resonant absorber. This introduces the second type of
experiment in which we allow the radiation to pass through a
resonant medium and then ask how the probability of detecting
a photon varies with time after the birth of the Mössbauer
14 keV state. We have shown that the exponential decay leads
to a Lorentz frequency distribution. It is natural to suppose
that if we modify the frequency distribution, the distribution
of observed decay times will not be exponential.

The principle of the calculation is straightforward although
the algebra becomes lengthy. We take the expression for the
decaying radiation, and obtain the Fourier transform for the
amplitudes, $a(\omega)$. We now represent the medium by a complex
refractive index, which will modify the $a(\omega)$ in both amplitude
and phase. We then take the Fourier transform again to obtain
the reconstructed probability distribution.

The medium is regarded as an assembly of resonators, and each
component $a(\omega)e^{i\omega t}$ produces forced oscillations of the resonators. The complex permittivity (complex dielectric constant)
may be written

$$\varepsilon(\omega) = 1 + r \left(\omega_o'^2 - \omega^2 + i\omega\Gamma\right)^{-1} \qquad (2.6)$$

We have written ω_o' as the resonant frequency to allow for a
difference between the source radiation and the resonant medium,
either because of hyperfine interactions or because of relative
motion introduced in the experiment. It can then be shown that
the effect of passage through the medium is to change $a(\omega)$ into
$a'(\omega)$, where

$$a'(\omega) = a(\omega) \exp \{-2ib\omega [\omega_o'^2 - \omega^2 + i \Gamma]^{-1}\} \qquad (2.7)$$

where b is a constant involving r and the thickness of the specimen.

Note that when $\omega = \omega_o'$, $a'(\omega'_o) = a(\omega_o') \exp (- 2b/\Gamma)$, so that the transmission on resonance is given by $\exp (-4b/\Gamma)$. The time dependence of the transmitted amplitude is then

$$a'(t) = \frac{1}{2\pi i} \int_{-\infty}^{+\infty} d\omega \frac{e^{i\omega t}}{\omega-\omega_o-1/2\ i\Gamma} \exp \frac{2ib\omega}{\omega^2-\omega_o'^2 -i\omega\Gamma}$$

$$(2.8)$$

The evaluation of this integral shows that the intensity is oscillatory in time. This implies the curious result that at certain times, the intensity through the absorbing medium is greater than when the absorber is removed. One may say that the medium is caused to "ring" by the incident radiation, so that beats are observed. Energy is stored during certain times and released at others. This interesting phenomenon has been observed experimentally. (Lynch et al, 1960).

Frequency Modulation

The radiation can readily be subjected to frequency modulation, (Lynch et al, 1960, Ruby and Bolef, 1960, Cranshaw and Reivari, 1967) and in fact the theory has already been given in Chapter 1. We there showed that when the emitting nucleus was in vibration with amplitude x_o and frequency ω sidebands appeared at frequencies $\omega_o \pm n\omega$, where n = 1,2 ..., with amplitude J_n^2 ($\kappa\ x_o$). This effect has been demonstrated by mounting either the source or the absorber on a piezo-electric crystal driven by an oscillator. A typical spectrum is shown in Fig. 2.1. An alternative way of looking at this phenomenon parallels the quantum mechanical approach to the theory of the Mössbauer effect in Chapter 1. We regard the mechanical vibration of the specimen as equivalent to putting a sharp peak in the phonon spectrum at the vibration frequency. Then emission of the Mossbauer γ-ray can be accompanied by the emission or absorption of n phonons, leading to a spectrum consisting of a

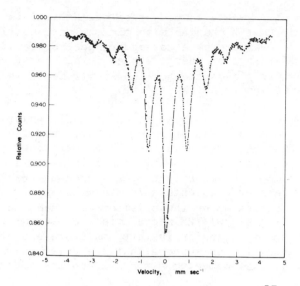

Fig. 2.1 The emission spectrum of a ^{57}Co source
vibrated at a frequency of 8 MHz

central line with equally spaced sidebands.

Phase Modulation and Amplitude Modulation

The phase modulation of the radiation has also been demonstrated.
(Grodzins and Phillips, 1961) Let us suppose that a length ℓ
of material of refractive index n be inserted in the path of
the radiation. Then Eq. 2.1 may be rewritten

$$\xi(x,t) = \xi_o \exp \{ i(\omega_o t - \kappa(x-\ell + \ell/n)) - \Gamma t/2 \}$$

$$= \xi_o \exp \{ i(\omega_o t - \kappa x - \emptyset) - \Gamma t/2 \} \qquad (2.9)$$

If ℓ changes, the phase angle $\emptyset$ changes, and the instantaneous
frequency observed is $\omega_o + \partial\phi/\partial t$

i.e. $\delta\omega = (1-n)\dot{\ell}$

$\delta\omega/\omega_o = (1-n)\dot{\ell}/c$ $\qquad (2.10)$

In the experiment, a wedge shaped piece of material is moved transversely so that ℓ changes with time. The resulting shift in energy can be measured by a Mössbauer spectrometer. The value of $(1-n)$ for rays of energy 14 keV is of the order of 10^{-6}, but the energy resolution of the spectrometer is good enough to measure $1-n$ to a few percent.

This experiment can also be interpreted in a different way. Consider a beam of radiation of energy density ε deflected by a prism or diffracted by an edge through a small angle θ. Then according to relativity theory, there is associated with the radiation a momentum density ε/c, and when the radiation is deflected, transverse momentum $\varepsilon/c \sin \theta$ is imparted to it. This momentum is taken up by the prism or edge and the energy transferred is $(\varepsilon/c \sin \theta)^2/2M$ where M is the mass of the prism, etc. Thus the energy transfer is very small and essentially zero. Now however, let us suppose that the prism is moving with velocity, v, i.e. momentum $p = Mv$. Then the energy is

$$E = p^2/2M$$

and $\delta E = p\ \delta p/M. = v\ \delta p.$

As before, if v=0, $\delta E = 0$. But if v has a finite value, $\delta E = v\delta p$ $= v/c\ \varepsilon \sin \theta$, and this energy is exchanged between the radiation and the prism. Now according to quantum theory, ε is proportional to frequency, i.e.

$$\frac{\delta E}{\varepsilon} = \frac{\delta \omega}{\omega_o} = (v/c)\sin \theta$$

Now for the deflection, θ, produced by a narrow angle prism of angle α we have $(1-n)\alpha = \sin \theta$

$$\therefore \quad \frac{\delta \omega}{\omega_o} = v\alpha(1-n)/c$$

which is the same result as before.

Notice that it is of no consequence how the deflection is brought about. It could equally well be a grating instead of a prism. If the spacing of the lines of the grating is d, then the deflection of the nth order of radiation of wavelength λ is

$$\theta = n\lambda/d$$

Thus the energy transfer will be

$$\frac{\delta\omega}{\omega_o} = \frac{v}{c} \ n\lambda/d \qquad (2.11)$$

or, if $f = \frac{\omega}{2\pi}$ is the frequency,

$$\delta f = nv/d \qquad (2.12)$$

i.e. the radiation has sidebands at $f \pm nv/d$. Now v/d is clearly exactly the "chopping frequency" if we regard the grating as chopping the incident radiation. (Isaak and Preikschat, 1972). If we regard this as modulating ω_o in equation (2.1), then we would find that the sideband intensities are given by the Fourier transform of the modulation. As is well known, the intensities of the orders in the diffraction by a grating are also given by the Fourier transform of the grating. The results by the two methods are therefore identical. It seems to be possible to imagine experiments in which one or the other method might provide the more appropriate description.

It is interesting to note the effect of passing the radiation through an inhomogeneous medium, moving transversely to the direction of the radiation. The energy distribution of the emerging radiation is broadened, and the observed distribution is related to the Fourier transform of the inhomogeneities of the material. The effect has been observed in such material as cardboard. (Champeney and Woodhams, 1966).

Faraday Effect

Finally, we briefly describe experiments in which the Faraday effect has been demonstrated for Mössbauer γ-rays. (Imbert,

1964, 1966) Consider apparatus arranged as in Fig. 2.2

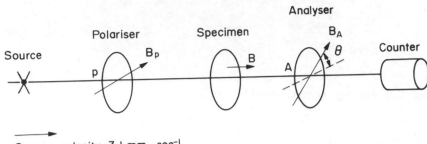

Source velocity 3.1 mm sec^{-1}

Fig. 2.2 The experimental layout of Mössbauer polarimeter

S is a source of ^{57}Fe radiation. P is a polarizer. This
consists of a thin foil of ^{57}Fe, with an applied magnetic field
B_p, in the plane of the foil and normal to the direction of
propagation of the γ-ray, and making an angle θ to an
arbitrary fixed axis. A is an analyzer, applied field B_A,
similar to the polarizer, but with the provision for the angle
$θ_A$ to be changed. If the source is now given an appropriate
velocity, the energy of the γ-rays passing through the
polarizer and analyser can be made to coincide with one of the
resonant transitions in the iron, say one of the $Δm = 0$
transitions. Then almost all the radiation polarized perpendicular to B_p will be absorber in the polarizer, and only the
component polarized parallel to B_p transmitted. If $θ_A = θ_p$,
clearly this radiation will also be transmitted by the
analyzer, and counted in the detector D. If $θ_A$ is varied, the
intensity recorded is approximately a sine function of $2∅$
where $∅ = θ_A - θ_p$. The apparatus therefore acts as a polarimeter. If $θ_A$ is varied and the count rate noted, any rotation
in the plane of polarization of the γ-ray can be measured as a
change of phase of the sine function.

Consider the case of a sample of iron foil, magnetized in the
direction of the γ-ray. (This would require a large field, of
about 2.1 T, to overcome the demagnetizing field. Alternatively,
nearly the same effect can be obtained by magnetizing the foil
in the plane of the foil, and then passing the γ-rays through
it at a large angle of incidence). Then the $Δm=0$ transitions
are forbidden, and the $Δm = \pm 1$, transitions lines 1,3,4,6 of
the iron spectrum are circularly polarized, left and right
handed as shown in Fig. 2.3, a. If the iron foil is at rest

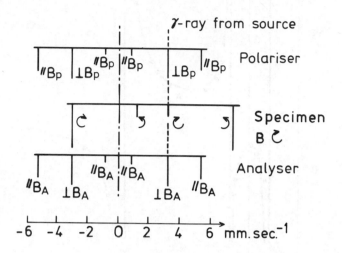

Fig. 2.3 a) Diagram showing the energy shifts and polarizations in the polarimeter of Fig. 2.2

relative to the polarizer and analyzer, the energy of the γ-ray which has been chosen to coincide with one of the $\Delta m=0$ transition in the polarizer and analyzer will clearly be resonant with the forbidden $\Delta m=0$ transition in the iron foil, and there will be no interaction. A small rotation of the plane will be observed due to the "tails" of the resonances at allowed transitions. However, if the foil is also moved with appropriate velocities, the γ-ray energy can be made to coincide with the $\Delta m = \pm 1$ transitions, and the rotation of the plane of polarisation produced by these transitions can be measured. The effect is shown in Fig. 2.3.b.

According to the classical theory of the Faraday Effect, the rotation angle produced by a transition of energy E_0 is given by

$$\delta = \text{const.} \quad \frac{(E_o - E_s)}{(E_o - E_s)^2 + \Gamma^2/4} \tag{2.13}$$

where E_s is the energy of the source γ-ray (with the appropriate

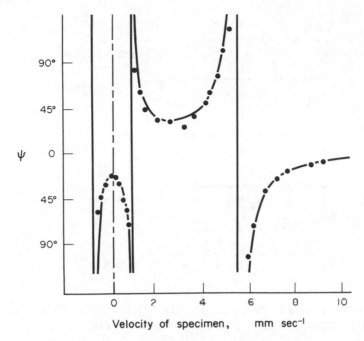

Fig. 2.3 b) Rotation of the plane of polarization by an iron foil magnetized along the γ-ray direction.

Doppler shift), Γ is the width of the resonance, and the constant depends on such quantities as the cross section for absorption, and the number of atoms of ^{57}Fe in the foil. The solid curve in fig. 2.3 b is calculated from this formula.

REFERENCES

Champeney, D.C. and Woodhams, F.W.D. (1966). Phys. Lett. 20 275.

Cranshaw, T.E. and Reivari, P. (1967). Proc. Phys. Soc. 90 1059.

Grodzins, L. and Philips, E.A. (1961). Phys. Rev. 124 774.

Imbert, P. (1964). Phys. Lett. 8 95.

Isaak, G.R. and Preikschat, E. (1972). Phys. Lett. 38A 257.

Lynch, F.J., Holland, R.E. and Hamermesh, M. (1960). Phys. Rev. 120 513.

Ruby, S.L. and Bolef, D.I. (1960). Phys. Rev. Lett. 5 5.

CHAPTER III

MÖSSBAUER EXPERIMENTS ON RELATIVITY

Some of the earliest experiments making use of the Mössbauer Effect were designed to test certain predictions made by relativity theory concerning the behaviour of clocks, or frequency standards which can be used as clocks. The Mössbauer effect permits us to use a natural frequency of vibration of an atomic nucleus as the standard.

Two things are required of a vibrating system if it is to be a good frequency standard. First, once excited, the rate of loss of energy must be low. Secondly the dependence of the natural frequency on the environment, e.g. temperature, pressure etc. must be low. We consider these in turn.

In macroscopic systems, the energy of excitation is usually lost by friction. For example, a balance wheel loses energy at the bearings, a tuning fork or quartz crystal loses energy by internal friction in the material, and an electrical tuned circuit loses energy by ohmic losses in the conductors. A smaller amount of energy, particularly in the case of the tuning fork or the tuned circuit may be radiated into the surroundings.

The rate of loss of energy is measured by Q, the number of cycles which must elapse before the energy has fallen to $1/e$ of its original value. Clearly, in order that a large number of cycles may be counted, Q must be large, and the accuracy to which the frequency is determined is given by $1/Q$. This conclusion is not invalidated if some device is arranged to keep the oscillations going, that is, to supply more energy at the rate at which it is being lost, because then if Q is small, the coupling to the device must be high, and variations in it will affect the frequency. Quartz crystals are amongst the best macroscopic vibrating systems, and for them Q may have the value $10^8 - 10^9$.

A nucleus is a bound system of protons and neutrons possessing excited states. The excitation energy from low lying levels is lost by either emission of electromagnetic radiation, that is γ-radiation, or by "internal conversion", the emission of an atomic electron, and the subsequent reordering of the electrons. In either case, the coupling is weak, and the Q relatively high. If the lifetime of the state is τ, and the energy $E = h\nu$, then $Q = \tau\nu$. We have shown in Chapter 1 that the lifetime τ leads to a Lorentz line shape for the emitted radiation of width $\Gamma = h/\tau$, so that an alternative interpretation of $1/Q$ is the relative width of the radiation spectrum. An inspection of nuclear tables will show that there is almost no limit to the values of Q that can be found. However, as explained in Chapter 1, the radiation from most of these nuclei will in practice be very much "broadened" by Doppler effects. It is only through the existence of the Mössbauer Effect that a practical source of very narrow line radiation can be produced, and also only through the resonance effect that very small changes in its frequency can be measured. The fact that the nucleus must have suitable values of parameters for the Mössbauer effect to be large greatly narrows the range of nuclei which can be considered, but even so, the values of Q which are obtainable show a great improvement over macroscopic systems. For example for ^{57}Fe, the 14 keV level has a Q of about 10^{12}.

The measurement of the shift of a line is best made by measuring the absorption on the sides of the absorption line, where the count rate changes most rapidly with shift. The precision of the measurement is then approximately $\Gamma n^{-1/2} \, d^{-1}$ where n is the number of counts and d the depth of the absorption.

We must now consider how the environment, and unavoidable changes in it may change the energy of the emitted γ-ray. Two effects have to be taken into account. The first is the temperature coefficient. This is itself a relativistic effect, and is described later. The second is the interaction of the nucleus with its atomic electrons, and charges on the nearby atoms, called hyperfine interaction and described in Chapter 5. (These interactions may also be temperature dependent, but the shifts so caused are usually smaller than the relativistic effect).

Provided that the material containing the Mössbauer nuclei is chemically stable, the hyperfine interactions will be constant

in time, and will not affect any measurements of shifts.
However, it may happen that the material is not homogeneous,
so that the Mössbauer nuclei are not all in exactly the same
environment, and the hyperfine interactions are then not
constant in space. This has the effect of broadening the
observed line.

Other important factors to be considered in choosing a suitable
nucleus are the "Mössbauer parameters", i.e. the value of f,
the recoilless fraction, and the size of the cross section for
resonance absorption. These determine the fraction of
emitted photons which can be absorbed on resonance, and
therefore the fraction of counts which are effective in determining a shift of frequency. One might also have to consider
the availability of a parent nucleus from which strong sources
of the radiation can be prepared.

Thus the ideal nucleus for measuring relativistic effects
should have (i) high Q, (ii) good Mössbauer parameters, i.e.
large recoilless fraction, f, large cross section, σ, (iii)
relative insensitivity to hyperfine interactions.

Three nuclei which might be considered are ^{57}Fe, ^{67}Zn and ^{181}Ta.
The relevant parameters are given in Table 3.1.

TABLE 3.1

Nucleus	E_o/keV	τ/s	W/eV	Q	depth of absorption	dQ
^{57}Fe	14.4	10^{-7}	$4.7 \ 10^{-9}$	$3 \ 10^{12}$	.30 at RT	$9 \ 10^{11}$
^{67}Zn	93	$9.4 \ 10^{-6}$	$5.0 \ 10^{-11}$	$2 \ 10^{15}$	.01 at 4°K	$2 \ 10^{13}$
^{181}Ta	6	$7 \ 10^{-6}$	$6.7 \ 10^{-11}$	10^{14}	.3	$3 \ 10^{13}$

In the sixth column, we have written the depth of absorption
which can be obtained. For the two nuclei ^{57}Fe and ^{181}Ta, we
have assumed an absorption depth of 0.3. Although deeper
absorptions can be obtained, they are inevitably accompanied
by line broadening because of the exponential character of the
absorption.

Most work has been carried out using the isotope [57]Fe, which permits strong absorptions at room temperature, and experiments with sources and absorbers on rotors are fairly straightforward. In the case of [181]Ta, the great sensitivity to hyperfine interactions has so far prevented the production of sources and absorbers with the natural line width. Recently, the technical difficulties of using [67]Zn have been overcome (Katila & Riski, 1981) and very striking results obtained, measuring the gravitational shift in as short a distance as 1 m.

We preface the discussion of the experiments with a resume of the relativistic results which lead to the predictions under test.

Relativistic Kinematics

During the first decade of the twentieth century, the principles of relativistic kinematics were laid down in a series of papers mainly by H.A. Lorentz, A. Einstein and H. Minkowski. In particular, in a paper entitled "On the electrodynamics of moving bodies" Einstein showed that relativistic laws of transformation of axes led to the conclusion that a moving clock loses time relative to a clock which is stationary. If a clock moves with constant speed v in a closed curve for time t, then by a clock which has remained at rest, the travelled clock will be slow by an amount $\frac{1}{2} t v^2/c^2$. (Einstein, 1905a). In the same paper he showed that the frequency ν' observed by an observer moving with velocity v at an angle θ with the direction of a distant source of light of frequency ν is given by

$$\nu' = \nu \frac{1 - \cos \theta \, v/c}{\sqrt{1 - v^2/c^2}} \qquad (3.1)$$

Further, by considering the surface enclosing a given quantity of radiation, he showed that the energy of radiation E' measured by the moving observer was given by

$$E' = E \frac{1 - \cos \theta \, v/c}{\sqrt{1 - v^2/c^2}} \qquad (3.2)$$

and comments that it is remarkable that the energy and the frequency of a light element vary with the state of motion of the observer in accordance with the same law.

We note that when $\theta = o$, the equations take the simple form

$$\nu' = \nu \; \frac{1 - v/c}{1 + v/c} \sim \nu(1 - v/c) \text{ for small } v \qquad (3.3)$$

and that when $\theta = \pi/2$

$$\nu' = \nu \; \frac{1}{\sqrt{1 - v^2/c^2}} \qquad (3.4)$$

the so-called "transverse Doppler effect". In a later paper the same year, (Einstein 1905b) Einstein showed that when a body emits radiation of energy E, its mass is reduced by the quantity E/c^2, leading to the idea of the equivalence of mass and energy, connected by the law

$$E = mc^2 \qquad (3.5)$$

It had been established in 1891 by R.V. Eotvos (Eotvos 1891) in a series of careful experiments that the "inertial mass" the quantity determining the acceleration of a body under a given force was equal to the "gravitational mass", the quantity determining the force existing between two bodies to within one part in 10^9. This is, in fact a refinement of Galileo's observation that all bodies fall equally fast, and the test has since been improved still further, (Dicke 1961) to a limit of less than one part in 10^{11}. In 1911, Einstein (Einstein 1911) put forward the Principle of Equivalence as an hypothesis without which the apparent identity of inertial and gravitational mass is accidental.

The Principle may be stated as follows:- a uniform gravitational field of strength γ along the z-axis is physically equivalent to a system in which there is no gravitational field but the coordinate axes are accelerated such that $\dfrac{\partial^2 z}{\partial t^2} = - \gamma$.

Then using this principle and the result for the transformation of energy, Einstein deduced that a gravitational field acts on the mass E/c^2 of radiated energy. For let a quantity E of energy be radiated from A to B along the z-axis a distance s, in a system under uniform acceleration. Then the time taken is to first order s/c. We may without loss of generality suppose that the velocity at A is zero at the instant of emission. Then the velocity at B is

$$- \frac{\partial^2 z}{\partial t^2} \, s/c,$$

and using the transformation law, the energy at B is

$$E' = E(1 - \frac{\partial^2 z}{\partial t^2} \, s/c^2) \qquad (3.6)$$

Then using the Principle of Equivalence, in a system in which a gravitational field acts, we must have $E' = E(1 + \gamma \, s/c^2)$. If the radiation moves from a point of gravitational potential $\emptyset_A$ to a point of gravitational potential $\emptyset_B$, there is a gain of energy E/c^2 ($\emptyset_A - \emptyset_B$), just as would be obtained by the gravitational field acting on a mass E/c^2. This result is, of course, entirely consistent with the result of Eotvos' experiments, and Einstein shows that it is also consistent with the conservation of Energy. From a modern standpoint it is natural to think in terms of photons and excited atoms, and we now reproduce the argument in these terms.

Let a series of atoms be moved from B to A and back to B. Let the atoms have masses M and excited states of mass M*. Let the atoms at B emit their energy in the form of a photon to an atom at A. Then the atoms that fall from B to A give up energy M* γS to the system and atoms that are raised from A to B take energy MγS. The conservation of energy requires that the photon of energy (M* – M)c loses energy (M* – M)γS on passing from A to B, i.e. a photon of energy E loses energy E/c^2 γS, the same quantity as deduced from the Principle of Equivalence. A photon of frequency ν at B has a frequency $\nu' = (1 - \gamma s/c^2)$ at A,

$$\text{or } \frac{\delta\nu}{\nu} = \frac{1}{c^2} (\emptyset_A - \emptyset_B) \qquad (3.7)$$

the so-called gravitational red shift.

Mössbauer Experiments

The magnitude of the gravitational red shift in the earth's
field is easily calculated to be

$$\frac{\delta\nu}{\nu} = g/c^2 \sim (9.81/9.10^{16} \sim 10^{-16} \text{ m}^{-1}.$$

If a height of 10m is available, then the shift is one part in
10^{15}. For the resonance in ^{57}Fe, Q is 10^{12} and the expected
shift is 10^{-3} line widths. From the order of magnitude argu-
ments presented in the introduction, a measurement with an
accuracy of 1% could be made by counting about 10^{11} photons.

It is interesting to note that for a given source, the count
rate is inversely proportional to the square of the height, and
the relative accuracy is thus independent of height. However,
at small heights the counting rate would become too high for
the counting equipment, and the red shift itself would be
small compared with the disturbing effects noted in the introduction.

A second experiment which has been carried out to test relativistic laws is concerned with rotating systems. The apparatus
in the simplest form of the experiment consists of a rotor
with a source mounted on the axis and an absorber mounted at
the circumference. The distance between the source and absorber
is thus constant, and no first order Doppler effect exists.
In a practical experiment, the source is distributed near
the axis and the absorber covers a finite length of the circumference. Indeed, in some experiments the source has been
mounted at the circumference at the same radius as the absorber.
It is convenient therefore to calculate the effects to be
expected generally. This may be done in several ways. We may
use 1) the formula for the relativistic Doppler effect, 2) the
formula for the time dilatation in a moving frame, or 3) we may
replace the force field in the rotating system by the gradient
of a potential, and use the same argument as before to deduce
the shift. Using the Doppler formulae, we imagine an absorber
moving with velocity v_a making an angle θ_a with the direction
of the radiation. Then by eq.(3.1) we find

$$\frac{\nu_a}{\nu_s} = \frac{1 - \cos\theta_a \, v_a/c}{\sqrt{1 - v_a^2/c^2}}$$

Similarly if an emitter moves with velocity v_e at an angle θ_e the frequency seen by a stationary observer is given by

$$\frac{\nu_s}{\nu_e} = \sqrt{\frac{1 - v_e^2/c^2}{1 - \cos \theta_e \, v_e/c}}$$

Then eliminating ν_s we have

$$\frac{\nu_a}{\nu_e} = \sqrt{\frac{1 - v_e^2/c^2}{1 - v_a^2/c^2}} \quad \frac{1 - \cos \theta_a \, v_a/c}{1 - \cos \theta_e \, v_c/c} \tag{3.8}$$

If $\theta_a = \theta_e = \pi/2$, eq. 3.8 reduces to

$$\frac{\nu_a}{\nu_e} = \frac{\sqrt{1 - \omega^2 r_e^2/c^2}}{\sqrt{1 - \omega^2 r_a^2/c^2}} \tag{3.9}$$

which is the result for the simplest case. However it is easy to see by geometry that the second term in the R.H.S. of (3.8) is unity for arbitrary points at r_e and r_a on the disc. Thus the ratio of the frequencies for extended emitters and absorbers is given purely by the time dilatation factors.

A rather direct way of obtaining the result is to replace the force field in the rotating system by the gradient of a potential $\emptyset$. Then

$$\frac{\partial \emptyset}{\partial r} = - \omega^2 r$$

and $\quad \emptyset_1 - \emptyset_2 = - \frac{\omega^2}{2} (r_1^2 - r_2^2)$

and using 3.7, $\frac{\delta v}{v} = \frac{1}{2} \frac{\omega^2}{c^2} (r_2^2 - r_1^2)$

However, it is important to note that this derivation makes no statements about gravitational fields, and the rotor experiment alone is in no sense a test of the Principle of Equivalence just because we may choose to derive the result in this way. We have tested only eq.(3.8) and the assumption that the force field in the rotating system is conservative.

A third measurement involving the Mössbauer effect also tests some aspects of relativity theory. This is the measurement of the temperature coefficient. As in the other two cases, there are several ways of deriving the result. The first, used by Josephson (1960) in a paper predicting the existence of the effect depends only on Einstein's conclusion that when a body emits a quantity of energy E, its mass is reduced by a quantity E/c^2. The derivation can be most easily seen for the Einstein model of the crystal. Here, each atom is a harmonic oscillator and has energy levels spaced by $h\nu$, and at the instant of emission of the γ-ray may have the quantum number n, and energy $nh\nu$. After emission of a Mössbauer γ-ray, the energy will be $nh\nu'$, where ν' is the new oscillator frequency.

Clearly $\nu'/\nu = \sqrt{M/M'}$ where $M-M' = E/c^2$

Thus the energy of the atom has increased by

$$\delta E = nh\,(\nu' - \nu) = -\tfrac{1}{2}\,\frac{M'-M}{M}\;nh\nu = \frac{E}{2Mc^2}\;nh\nu$$

and this energy is taken from the γ-ray. $nh\nu = U$ is the energy of the oscillator at temperature θ. Thus

$$\frac{\delta E}{E} = \frac{U}{2Mc^2} \tag{3.10}$$

and

$$\frac{\delta E}{E} \cdot \frac{1}{\delta\theta} = \frac{dU}{d\theta}\,\frac{1}{M.2c^2} = \frac{C_p}{2c^2} \tag{3.11}$$

where C_p is the specific heat. We may also derive (Pound, 1960) the result by noting that in the Mössbauer transition, the mean value of the velocity of the atom is zero. However the mean square velocity is not zero, but is given by

$$< \tfrac{1}{2} Mv^2> = K, \text{ the kinetic energy, or}$$

$$<v^2> = 2K/M = U/M$$

Thus clocks moving with the nucleus suffer a time dilatation $1 - \tfrac{1}{2}<v^2>/c^2$ so that the frequency seen in the rest frame is changed by

$$\frac{\delta\nu}{\nu} \sim \tfrac{1}{2} \frac{<v^2>}{c^2} = \frac{U}{2Mc^2}$$

leading to 3.11 as before.

In this derivation, we have assumed the time dilatation formula and implicitly assumed that accelerations have no effect. As a test of the time dilatation formula, the rotor experiments may be preferred because other factors also shift the frequency with temperature, and these are not easily calculated. However, it is worth noticing that the accelerations involved are extremely large. For example, we may estimate the frequency of the atom due to lattice vibrations from $h\nu \sim k\theta_D$ where θ_D is the Debye temperature. Then $\nu \sim 10^{14}$ cps, and the acceleration is of the order 10^{15} g. whereas acceleration in the rotor experiments is of the order 10^5 g. only. The measurements of the temperature shift and their adequate explanation in terms of the second order Doppler effect shows that no changes of frequency greater than a few parts in 10^{12} are produced by accelerations of the order 10^{15} g.

REFERENCES

Dicke, R.H.   (1961) Sci. Am. 205 84.
Einstein, A. (1905) Ann. der Physik. 17 891.
Einstein, A. (1905 b). Ann. der Physik, 17.
Einstein, A. (1911). Ann. der Physik, 35 898.
Eotvos, R.V. (1891). Math. Natur Berichte Ungarn 8 65.
Josephson, B.D, (1960). Phys. Rev. Lett. 4 341.
Katila, T. and Riski, K.J. (1981) Phys. Rev. Lett. 83A 51.
Pound, R.V. (1960) Phys. Rev. Lett. 4 274.

CHAPTER IV

LATTICE DYNAMICS

The atoms in a solid are vibrating. The traditional way of obtaining information about the vibrations is from specific heat measurements, especially at low temperatutes. The interpretation of these by Einstein and by Debye was a triumph for the theory in the early days of quantum physics. More modern methods use thermal neutron scattering to measure the frequency spectrum of the vibrations.

The Mössbauer Effect offers two separate ways of studying atomic motion in solids:

(i)   from data on the intensity f of the Mössbauer Effect. This is given by

$$f = \exp\left(- \kappa^2 \langle x^2 \rangle\right) \tag{4.1}$$

and so gives a measure of the mean square displacement $\langle x^2 \rangle$ of the atoms along the direction (x) of the γ-rays.

(ii) from the second order Doppler shift of the Mössbauer spectrum measured as a function of temperature. This is from Eq. (3.12).

$$\delta_T \ \sim \ -\tfrac{1}{2}\ \frac{\langle v^2 \rangle}{c} \tag{4.2}$$

and so measures the mean square velocity of the atoms.

Classical Theory

The early theories of vibrations in solids considered the atoms to behave as independent classical three-dimensional harmonic oscillators. We need to find the average displacement of the atoms from their equilibrium positions and their average velocity.

Consider first one-dimensional motion for simplicity. The displacement of the atoms from their equilibrium lattice sites may be written

$$x = x_o \cos \omega t \tag{4.3}$$

where x_o is the amplitude and ω the frequency of the vibration. Thus the velocity and acceleration at time t are given by

$$\dot{x} = - x_o \omega \sin \omega t \tag{4.4}$$

$$\ddot{x} = - x_o \omega^2 \cos \omega t \tag{4.5}$$

For a monatomic solid, if λ is the force constant between atoms of mass M, the equation of motion is

$$M\ddot{x} = - \lambda x \tag{4.6}$$

From 4.3 and 4.6

$$\omega^2 = \frac{\lambda}{M} \tag{4.7}$$

The kinetic and potential energies are

$$K = \tfrac{1}{2} M \dot{x}^2 = \tfrac{1}{2} M x_o^2 \omega^2 \sin^2 \omega t \tag{4.8}$$

$$P = \tfrac{1}{2} \lambda x^2 = \tfrac{1}{2} \lambda x_o^2 \cos^2 \omega t$$

$$= \tfrac{1}{2} M x_o^2 \omega^2 \cos^2 \omega t \tag{4.9}$$

by (4.7), so that the total vibrational energy of an atom is

$$K + P = \tfrac{1}{2} M x_o^2 \omega^2 \tag{4.10}$$

independent of time. On average this must be equal to the
thermal energy of a one-dimensional oscillator, which may be
shown to be kT by integrating over a Boltzmann distribution
of energies (the law of equipartition of energy). So

$$\tfrac{1}{2} M x_o^2 \omega^2 = kT \tag{4.11}$$

i.e.

$$x_o = \sqrt{\frac{2kT}{M\omega^2}} \tag{4.12}$$

The mean square displacement is then

$$\langle x^2 \rangle = \langle x_o^2 \cos^2 \omega t \rangle$$

$$= \tfrac{1}{2} x_o^2$$

$$= \frac{kT}{M\omega^2} \tag{4.13}$$

(this is a quantity which also appears in the classical theory
to explain high temperature behaviour of electrical resistivity
of metals). The mean square velocity is

$$\langle \dot{x}^2 \rangle = x_o^2 \omega^2 \sin^2 \omega t$$

$$= \tfrac{1}{2} x_o^2 \omega^2$$

$$= \frac{kT}{M} \tag{4.14}$$

For the second order Doppler shift we need the three-dimensional
value of the mean square velocity which for an isotropic solid
is just three times the one-dimensional value

$$\langle v^2 \rangle \; = \; 3 \; \langle \dot{x}^2 \rangle \; = \; \frac{3kT}{M} \qquad\qquad (4.15)$$

So that at high temperatures, where classical theory is valid, f and δ_T are given by

$$f \; = \; \exp \left(- \kappa^2 \; \frac{kT}{M\omega^2} \right)$$

and

$$(4.16)$$

$$\delta_T \; = \; - \frac{3}{2} \; \frac{kT}{Mc}$$

We may note that for a mole of a solid the thermal energy follows from 4.11 to be

$$U \; = \; 3 \; N \; kT \qquad\qquad (4.17)$$

when N = Avogadro's number and the factor 3 comes from the three degrees of freedom of the motion. Thus the specific heat is

$$C_v \; = \; \frac{\partial U}{\partial T} \; = \; 3Nk \qquad\qquad (4.18)$$

i.e. it is independent of temperature. This we recall is in agreement with experimental data at high temperatures, but at low temperatures the specific heat of solids falls towards zero.

Quantum Theory

(i) Einstein theory

Classical theory failed to explain the specific heat of solids at low temperatures, although it was quite successful for many solids at high temperatures. The discrepancy arises because the oscillations are quantised, which has a profound effect on their behaviour. A quantum mechanical harmonic oscillator cannot have any arbitrary value for its energy, but has energy eigenvalues which are equally spaced by $\hbar\omega$ where ω is the frequency of oscillation. Furthermore the lowest energy state of the oscillator does not have zero energy, but has a zero-point energy of $\frac{1}{2}\hbar\omega$. The energy of a single oscillator is thus $(n + \frac{1}{2})\hbar\omega$ where $n = 0, 1, 2.....$ The average energy of an assembly of such oscillators is

$$\sum_{n=o}^{\infty} (n + \tfrac{1}{2})\hbar\omega \, e^{-(n + \frac{1}{2})\hbar\omega/kT} \Big/ \sum_{n=o}^{\infty} e^{-(n + \frac{1}{2})\hbar\omega/kT}$$

$$= (\bar{n}_{\omega} + \tfrac{1}{2})\hbar\omega \tag{4.19}$$

where

$$\bar{n}_{\omega} = \frac{1}{e^{\hbar\omega/kT} - 1} \tag{4.20}$$

is the Bose - Einstein distribution function.

Einstein's approach was to consider the solid to be described by 3N independent quantized oscillators of one frequency ω only. Thus the total energy of a mole of a solid is

$$U = 3N (\bar{n}_{\omega} + \tfrac{1}{2})\hbar\omega \tag{4.21}$$

and the specific heat is

$$C_{\nu} = \frac{\partial U}{\partial T} = \frac{3 \, Nk\left(\frac{\hbar\omega}{kT}\right)^2 e^{\hbar\omega/kT}}{(e^{\hbar\omega/kT} - 1)^2} \tag{4.22}$$

This gives a specific heat which falls to zero at 0 K. (Note that Einstein did not actually include the zero point term, but as this is a constant it does not affect the specific heat result).

The amplitude of the atomic displacements in quantum theory is given by equating the energy of an oscillator $\frac{1}{2} M \omega^2 x_0^2$ (or $M \omega^2 \langle x^2 \rangle$) to $(n_\omega + \frac{1}{2})\hbar\omega$. So

$$\langle x^2 \rangle = (\overline{n}_\omega + \tfrac{1}{2}) \frac{\hbar}{M\omega} \tag{4.23}$$

and

$$\langle v^2 \rangle = 3 (n_\omega + \tfrac{1}{2}) \frac{\hbar\omega}{M} \tag{4.24}$$

since $\langle v^2 \rangle = 3 \langle x^2 \rangle \omega^2$. Note that in the high temperature limit $\overline{n}_\omega \to kT/\hbar\omega$ and we can ignore the factor $\frac{1}{2}$ arising from zero-point motion, so that the classical results 4.15 and 4.16 are obtained.

(ii) <u>Debye Theory</u>

The Einstein theory explained the observed fall off in specific heat at low temperatures although the quantitative agreement with experiment was not good. To get better agreement it is necessary to take account of the fact that the coupling between adjacent atoms causes their motion to be correlated. The vibrations can be shown to be in the form of waves (phonons) with a frequency spectrum $Z(\omega)$, where $Z(\omega)$ is the number of modes with frequency between ω and $\omega + d\omega$.

The Debye model assumes that there is frequency cut-off at $\omega = \omega_D$ and that for $\omega < \omega_D$ the frequency distribution obeys the relation $Z(\omega) = A\omega^2$, where A is a constant.

A is found from the condition

$$\int_0^\infty Z(\omega) \, d\omega = 1 \tag{4.25}$$

with

$$\left. \begin{aligned} Z(\omega) &= A\omega^2 \quad \text{for} \quad \omega < \omega_D \\ &= 0 \quad\quad \text{for} \quad \omega > \omega_D \end{aligned} \right\} \tag{4.26}$$

This gives

$$A = 3/\omega_D{}^3 \qquad\qquad\qquad (4.27)$$

and the specific heat

$$C_\nu = 3Nk \int_0^{\omega_D} \left(\frac{\hbar\omega}{kT}\right)^2 \frac{e^{\hbar\omega/kT}}{(e^{\hbar\omega/kT} - 1)^2} Z(\omega) \; d\omega$$

$$= 9Nk \; \frac{T^3}{\Theta_D} \int_0^{\Theta_D/T} \frac{e^x x^4 \, dx}{(e^x - 1)^2} \qquad\qquad (4.28)$$

where $\Theta_D = \hbar\omega_D/k$ is the characteristic Debye temperature for the solid. This agrees with the classical theory result at high temperatures and with the observed T^3 behaviour at low temperatures.

For the Mössbauer recoilless fraction we need the mean value of $\hbar/M\omega$, averaged over the vibration frequencies, i.e.

$$\langle x^2 \rangle = \int_0^{\omega_D} (\bar{n}_\omega + \tfrac{1}{2}) \, Z(\omega) \frac{\hbar}{M\omega} \; d\omega \qquad\qquad (4.29)$$

We then obtain

$$f = \exp\left\{-\frac{3}{2} \; \frac{E_R}{k\Theta_D} \left[1 + 4\left(\frac{T}{\Theta_D}\right)^2 \int_0^{\Theta_D/T} \frac{x \, dx}{e^x - 1}\right]\right\}$$

$$\qquad\qquad (4.30)$$

where we have written $k\,\Theta_D = \hbar\omega$, $\kappa = E_\gamma/\hbar c$ and $E_R = E_\gamma^2/2Mc^2$

To find the second order Doppler Shift we must average $\hbar\omega/M$ for the solid, i.e.

$$<v^2> = \int_0^{\omega_D} (\bar{n}_\omega + \tfrac{1}{2})\, Z(\omega)\, \frac{\hbar\omega}{M}\, d\omega \qquad (4.31)$$

and hence

$$\delta_T = -\frac{9\,k\Theta_D}{Mc}\left\{ \frac{1}{4} + 2\left(\frac{T}{\Theta_D}\right)^4 \int_0^{\Theta_D/T} \frac{x^3\,dx}{e^x - 1} \right\} \qquad (4.32)$$

Measurements of f and δ as a function of temperature have been used in conjunction with 4.30 and 4.32 to yield information on the lattice dynamics of solids.

CHAPTER V

HYPERFINE INTERACTIONS

The hyperfine interactions may be regarded as those small additional static interactions which occur between a nucleus and its environment as a consequence of the fact that a nucleus is not a structureless point charge, but a cluster of moving charges, distributed over a finite volume. The Mössbauer effect affords a mechanism for measuring these small effects, and in so doing provides a highly detailed account of the way in which a nucleus interacts with its environment.

Consider the interaction between a charged body and the electrostatic field produced by a distribution of other charges. The total energy of the system can be considered to be the sum of a series of interactions:-

i) The Monopole Interaction: i.e. the interaction between the charges on the body and the electrostatic potential generated by charges outside the body.

ii) The Dipole Interaction: If the charged body has a dipole moment, one end of the dipole will seek regions of low electrostatic potential, the other end will seek regions of high potential: that is the dipole will interact with the first derivative of the electrostatic potential, known as the potential gradient or electric field. At the origin of coordinates the field generated by a charge ξ at the point, r, θ is $- \xi \cos \theta / 4\pi\epsilon_o r^2$. If the coordinates are defined with respect to a dipole of moment D, the dipole interaction is clearly $- D \xi \cos \theta / 4\pi\epsilon_o r^2$.

iii) The Quadrupole Interaction: The charged body may have a quadrupole moment. For our purposes a quadrupole may be regarded as two dipoles fixed end to end in opposition (See Fig. 5.1). Since one dipole will seek regions of positive field and the other will seek regions of negative field, interactions will occur only if the field is changing with respect to the the coordinates; that is, if there is a field gradient

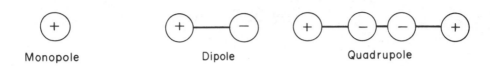

Fig. 5.1   Schematic illustration of a monopole,
a dipole and a quadrupole

present. A field gradient is the first derivative of the
electric field, the second derivative of the electric potential.
Some elementary calculus and trigonometry will show that a
charge ξ at the point r, θ generates an electric field
gradient $\xi\,(3\cos^2\theta-1)/4\pi\epsilon_o r^3$ along z at the origin of
coordinates. Hence the interaction between a quadrupole of
moment Q, at the origin with its axis along z, and a charge
at the point (r, θ) is $Q\,(3\cos^2\theta-1)/4\pi\epsilon_o r^3$. It should be
clear from Fig. 5.1 that unlike a dipole, a quadrupole can be
either positive or negative, depending upon which polarity
lies at the extremities. The charged body may also have
higher multipole moments, but these may be ignored, as they
lead to extremly weak interactions.

iv) <u>Magnetic Interactions</u> If the charges on the body are not
static, the possibility arises of magnetic moments being
generated. The only such effects that need concern us are when
charges on the body describe circular paths within the body
(generating orbital angular momentum) or when they appear to
be spinning about their own centres (generating so-called spin
angular momentum). In either case magnetic dipole moments will
be produced, and such dipoles will clearly interact with any
magnetic field that may be present, however produced.

Let us now apply these results to the interaction between a
nucleus and its electrons.

(i) The Monopole Interaction

The monopole interaction is the force which binds the electrons
to the nucleus and it makes the dominant contribution to the
total energy. Usually it can be described in terms of the
Coulomb interaction between point charges, but in considering
hyperfine interactions it is necessary to make a small correc-
tion to account for the finite size of the nucleus. It is
easy to see how this effect arises if we consider the energy
of the total system, electrons plus nucleus, and consider the
potential field generated by the nucleus. The potential
generated by a point charge varies as 1/r (r being the radius
vector), tending to infinity as r approaches zero. We can
approximate the effect of finite nuclear volume by considering
the nucleus to be a conducting, charged, hollow sphere, and
noting that in this case the potential varies as 1/r only in
the region outside the sphere. Inside the sphere the potential
is constant, $v = \xi/4\pi\epsilon_0 r_s$, where r_s is the radius of the
sphere and ξ is the charge. (We could alternatively consider
the nucleus to be a solid conducting charged sphere but the
essence of the result is the same: the potential is finite
at r = 0). Now, the wavefunctions of certain atomic electrons
(s electrons) have finite amplitudes at r=0. They therefore
have finite probabilities of being found inside the nucleus,
where, as we have seen, the potential is less than it would
have been if the nucleus had been a point charge. The interaction between nucleus and electrons is thereby diminished and
the energy of each atomic state is raised by a very small
amount relative to the corresponding state of a hypothetical
atom having a point charge nucleus. The magnitude of this
effect is dependent upon the effective radius of the nucleus and
the density of electrons within it. We have been talking here
about states of the whole atom. However, to a very close
approximation Mössbauer transition can be regarded as taking
place between two states of a nucleus, with the electronic
states unchanged, and the nuclear states then appear to have
had their energies changed by the nuclear volume effect.

If the size of a given nucleus remained constant, the volume
effect would be present, but there would be no means of
measuring it. However Mössbauer spectroscopy is concerned with
transitions between different nuclear states, and such
transitions will be accompanied by change in nuclear volume.
The energy of the Mössbauer transition contains a small term
corresponding to the difference between the nuclear volume

effects in the upper and lower states. Unfortunately, we still have no means of measuring this small term, partly because we have no reference energy, such as the energy of the transition in a point charge nucleus, and partly because we have no means of measuring the γ-ray energy to the required accuracy. The nuclear volume effect only becomes apparent when, in a Mössbauer experiment, the electron density at the nuclei in the source material is different from that in the absorber material. The Mössbauer transition energy will now be slightly different in the two materials and this will appear as a shift in the Mössbauer spectrum see Fig. 5.2a. This shift is known as the <u>isomer shift</u> or <u>chemical shift</u>, and is proportional to the difference in the electron densities at the nuclear sites in the two materials.

For a given Mössbauer isotope, measurements of the isomer shift allow us to observe how the electron density at the nucleus changes as we pass from one material to another, and this is extremely useful. We are not able to measure absolute values for differences in electron density (in the sense of electrons per unit volume) but this is not as serious as it sounds. The electron density at the nucleus is controlled by several different mechanisms, and the degree of complexity is such that an understanding of isomer shift trends is often as much as we can hope to achieve.

The first point to be considered is the total number of s electrons present. Changes of valency, or of electron configuration which involve changes in the total number of s electrons, will usually be accompanied by significant changes in the isomer shift. (It is important to remember, however, that valence shell s electrons are usually only a small proportion of the total s electrons present). Secondly, we need to consider the 'purity' of s electron orbitals. Any electron configuration that we assign to an atom is probably, at best, relevant only in the gas phase. When an atom becomes bonded into a solid its true wavefunction is unlikely to correspond, even approximately, to any single electron configuragion written in terms of atomic wave-functions. The s orbitals which are low-lying but empty in the gas phase atom may, in the solid, interact, and become mixed with occupied orbitals of the surrounding atoms, resulting in an apparent donation of electrons from the surrounding atoms into the vacant s orbitals. Similarly, electrons which occupy s orbitals in the gas phase atom may, by a similar mechanism, become extensively delocalized over surrounding atoms upon entering the solid. Furthermore, orbitals which are pure s in the high symmetry gas phase environment may, on entering the relatively low symmetry environment of a solid become mixed with p,d orbitals of the same atom. This will result in an apparent flow of electrons either into or out of s orbitals, depending upon how the orbitals are occupied in the gas phase atom. The third major factor influencing the isomer shift is probably the most troublesome. The wave-function of an s electron is not independent of the number and type of other electrons present in the same atom. The density of p, d, f, etc. electrons will influence the spatial distribution of s electrons on the same atom through inter-electronic repulsion, and the amplitudes of s electron wave-functions at the nucleus will therefore be dependent upon electron densities elsewhere in the atom. The p, d, f electrons are said to screen (or shield) the s electrons from the nucleus. Given that screening effects are comparable in importance to the other factors that have been mentioned, and that occupancy of p, d, f, orbitals are governed by complexities similar to those governing the occupancy of s orbitals the difficulties inherent in interpreting isomer shift results will be appreciated. Nevertheless many successful and important studies of isomer shift trends and fluctuations have been made, and these have contributed greatly to our understanding of electron behaviour in metals, alloys, chemical compounds etc. Specific examples are given in later chapters.

Running parallel to such qualitative studies there have been, inevitably, many attempts to extract absolute numbers from measurements of the isomer shift, in spite of the difficulties. Such work is useful and necessary as a test and check of current theories of quantum mechanics.

If we ignore relativistic effects, the isomer shift in a Mössbauer experiment is given by:-

$$\delta = (Z\, e^2\, R^2\, c/5\varepsilon_o E_\gamma) \left[\rho_a(0) - \rho_s(0) \right] \left[\Delta R/R \right] \text{ mm s}^{-1}$$

5.1

where Z is the atomic number, e is the electronic charge, R is the effective nuclear radius, c is the velocity of light, E_γ is the energy of the Mössbauer γ-ray, $\rho_a(0)$, $\rho_s(0)$ are the total electron densities at the nucleus for source and absorber respectively, and $\Delta R = R_{excited} - R_{ground}$. To a good approximation R is given by $R = 1.2\ A^{1/3}$ Fermis, where A is the nuclear mass number. We have, therefore

$$\delta = (1.56 \times 10^{-25}\ Z\ A^{2/3}/E_\gamma)\left[\rho_a(0) - \rho_s(0)\right]\left[\Delta R/R\right]\ \text{mm s}^{-1}$$

5.2

Equation 2 is written as a product of a constant which is known for any given Mössbauer transition, and the two terms in square brackets. There is no way of measuring either of the latter terms separately by experiment and the only available option is to estimate one by calculation. The current state of quantum mechanical calculations dictate that we proceed by calculating

$$\left[\rho_a(0) - \rho_s(0)\right]$$

for carefully selected source-absorber pairs, measure δ, and thus obtain a value for $\Delta R/R$. Estimates of $\Delta R/R$ by this method are not particularly reliable. In the most favourable case, ^{57}Fe, estimates of $\Delta R/R$ differed by a factor of 5 over a period of ten years, although they do now seem to be converging on a value of $- 0.8 \times 10^{-3}$. Corresponding values for other isotopes are even less certain, but continuous progress is being made.

(ii) The Dipole Interaction

No nucleus has ever been found to possess a static dipole moment. The reason for this is that the charge density ρ at any point is given by $\rho = \psi^*\psi$ where ψ is the total wave function (normalised) for charge-bearing nuclear particles, and ψ^* is its complex conjugate. ρ is therefore always an even function of the coordinates, and in a centro-symmetric system must be invariant with respect to inversion through the centre.

(iii) The Quadrupole Interaction

It is clear from our earlier illustrations that a quadrupole moment is quite compatible with a charge density which is symmetric with respect to inversion through the centre, and many nuclear states are known to have quadrupole moments. This doesn't imply that there are negative charges on the nucleus—merely that the nuclear charge need not be distributed spherically. An elongated positive charge can be considered as a positive monopole plus a positive quadrupole, a flattened charge as a positive monopole plus a negative quadrupole. The energies of the states are therefore influenced by the presence of electric field gradients (efg's). From the analysis given earlier, it will be apparent that an efg will be generated at the nucleus whenever the nuclear environment has a charge symmetry lower than cubic. Quadrupole interactions form a useful tool for investigating charge distributions in non-cubic materials.

The interaction between a nuclear quadrupole and an efg is quantised, and splits the nuclear sub-states without removing the degeneracy between states having the same value of M_I (Readers not familiar with Quantum Mechanics should see the 'Note on Quantisation' at the end of this chapter). The effect that this has on the Mössbauer spectrum can best be illustrated by taking the simplest known case, that of ^{57}Fe. Here the nuclear ground state has a spin of 1/2, while the state at 14.4 keV has a spin of 3/2. The presence of an efg does not effect the ground state, but in the upper state it removes the degeneracy between the $M_I = \pm 1/2$ and the $M_I = \pm 3/2$ pairs of substates. If the efg is positive in sign the effect is as shown in Fig. 5.2b.

It is easy to see why a positive efg splits the states as shown. A positive efg corresponds to a preponderance of negative charge in the x,y plane around the nucleus. The quadrupole moment is positive in the excited state of ^{57}Fe, and it is energetically more favourable for the quadrupole moment to lie near the x,y plane than to lie near the z axis. Remembering that the M_I values are the components of the angular momentum along the z axis it is clear that the $M_I = \pm 1/2$ states lie at lower energy than the $M_I = \pm 3/2$ states. (Note, however, that there is no way that the _sign_ of the efg can be obtained from the spectrum in Fig. 5.2b, since the quadrupole − split spectrum is symmetric about the centroid. For nuclei other than ^{57}Fe involving spins greater than I = 3/2 the quadrupole-split spectrum is more complicated, and the sign of the interaction can be seen by inspection). In the case of ^{57}Fe it is necessary to apply a magnetic field to the sample to ascertain the sign of the efg, or make a measurement

6

54

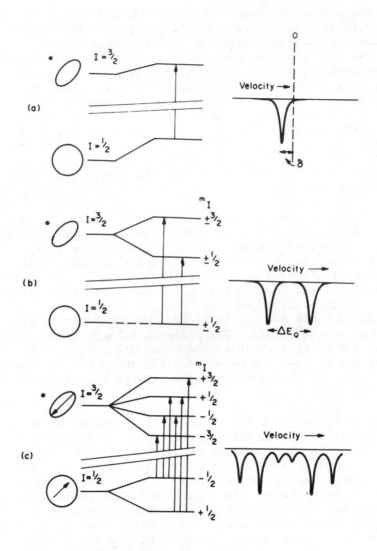

Fig. 5.2 Hyperfine interactions for the ^{57}Fe nucleus. On the extreme left of each diagram are shown the unperturbed nuclear levels, the I=3/2 state being 14.4 keV above the I=½ ground state. The effect of increasing the s-electron density at the nucleus is shown in a. Relative to the hypothetical point-charge nucleus, the energies of both states are increased, but the more voluminous ground state is affected most.

on a single crystal.

The sources of efg's at nuclear sites are basically of two kinds; those generated by charges in the lattice and those generated by the electronic charge cloud of the atom itself, although this distinction can often become blurred. With regard to the first - if we set a closed-shell atom at an octahedral site in a cubic lattice there will be no efg. If the charges along the Z axis were to move in towards the central atom, a negative efg would be generated, and a quadrupole splitting would appear in the Mössbauer spectrum. If the charges were to move out rather than in, a positive efg would be generated, and providing the sign of the interaction could be determined as described below, the nature and extent of such distortions could be studied. Efg's generated by the electron cloud are produced by open-shell electrons only (Closed shells have spherical symmetry. s-electrons do not contribute for the same reason). For example the electronic configuration of the Fe^{++} ion can be considered to be six 3d electrons outside a closed shell. In most cases the first five d electrons can be allocated, one to each of the five d orbitals. This arrangement has spherical symmetry, so it is only the sixth electron that need concern us. This electron will occupy the lowest lying d-orbital*, and will generate an efg at the nucleus given by the average value of $(3 \cos^2 \theta - 1)r^{-3}$ for the orbital in question. The efg's thus generated are usually substantially greater than those generated by the lattice charges, mainly because the r^3 term is smaller. This is fortunate, since as we indicate in the footnote[1], an electronic charge cloud will only generate an efg in a situation where the lattice would also be expected to be producing one.

If the Mössbauer atom is bonded covalently to its surroundings, its electrons will be delocalized over the adjacent lattice

---

[1] The symmetry must be sufficiently low for one d-orbital to be energetically favoured over the other four; otherwise the electron will average itself over the orbitals and retain spherical symmetry.

atoms and the distinction between efg's generated in the atom
and those generated in the lattice will be largely lost. The
quadrupole coupling is now giving information about the average
value of $(3 \cos^2 \theta - 1)/r^3$ for <u>molecular orbitals,</u> and if handled
competently can give useful information about the nature and
the extent of covalent bonding.

(iv) <u>Magnetic Interactions</u>

A Mössbauer-type transition inevitably involves a change in the
nuclear spin-state. It therefore follows that at least one,
and more usually both, of the nuclear states involved will
have non-zero spin. A strong enough magnetic field will
therefore always cause splittings in a Mössbauer spectrum.
(See note on Quantisation). The sort of effect that may be
observed can be seen in Fig. 5.2c where we show the effect of
magnetic perturbation upon the Mössbauer spectrum of the ^{57}Fe
nucleus, which again provides the simplest practical illustration. It will be noted that the upper state, with I = 3/2
splits into four, and that the ground state, with I = 1/2
splits into two. This might be expected to produce an eight-
peak spectrum. However, there is a rule that the value of M_I
must not change by more than one during a Mössbauer transition
and this will be seen to restrict the allowed transition to
six. The splittings between the peaks are directly proportional to the magmitude of the field at the nucleus. For the
case in which the magnetic field is perpendicular to the γ-ray
beam, the intensities of the transition are in the ratios
3:4:1:1:4:3. These ratios are readily established from very
simple principles (Kuhn, 1969, p211). When the field is
parallel to the γ-ray beam the ratios become 3:0:1:1:0:3, and
use is frequently made of the difference between the two
situations.

It is instructive to see how a magnetic field can be generated
at the nucleus. The case in which an external magnetic field
is applied to a diamagnetic material is easy - the field at
the nucleus is equal to the applied field. Cases where it is
useful to measure the Mössbauer spectrum of a diamagnet in an
applied field are considered in the next section. The more
interesting case arises when the atom itself has a magnetic
moment due to the presence of unpaired electrons. Two possibilities then arise: magnetically ordered materials, and
paramagnetic materials. In the first of these the unpaired
spins are coupled together throughout the material. If we

again take ^{57}Fe as our example, the unpaired electrons will be in the 3d orbitals. How do these electrons produce a magnetic field at the nucleus? The most important mechanism involves the exchange interaction, which has the effect of reducing slightly the repulsion between electrons having the same value of the magnetic spin quantum number m_s.[1] Most of the electrons in an iron atom exist in pairs (one having $m_s = +$ 1/2 the other $m_s = -$ 1/2, the net spin being zero). The unpaired electrons in the 3d orbitals will all have the same spin (say $m_z = +$ 1/2), and will 'attract' towards them all other electrons having $m_s = +$ 1/2. This will leave a preponderance of $m_s = -$ 1/2 electrons at all other points in the atom. This mechanism is particularly effective where it is the s electrons which are being polarized, since they have high densities at the nucleus, and interact with it by what is known as the Fermi contact interaction. The nett result is a large field, H_s, at the nucleus, opposite in sign to the field generated within the bulk of the material by the spins in the 3d orbitals.

$$H_s = -\ (16\pi/3)\ \mu_B\ \langle \Sigma(\phi_{ns,o}(+\ 1/2) - \phi_{ns,o}(-\ 1/2)\rangle,$$

where μ_B is the Bohr magneton and $\phi_{ns,o}(+\ 1/2)$, $\phi_{ns,o}(-\ 1/2)$ are, respectively, the s-electron positive and negative spin densities at the nucleus.

The second mechanism concerns orbital motion of the unpaired electrons about the nucleus. This motion constitutes a circular electric current and generates a magnetic field H_L at the nucleus

$$H_L = -\ 2\mu_B\ \langle r^{-3}\rangle\langle L\rangle.$$

Here L is the orbital angular momentum vector and the angled brackets, $\langle\ ,\ \rangle$, indicate integration over the whole atom.

Finally the unpaired electrons are essentially small bar magnets, due to their spin, and generate a magnetic field H_D at the nucleus by the normal dipole interaction.

$$H_D = -\ 2\mu_B\ \langle S\rangle\langle r^{-3}\rangle\ \ \langle 3\ \cos^2\theta - 1\rangle$$

where S is the spin angular momentum vector.

---

[1] We use lower case letters for quantum numbers of particles, upper case letters for quantum numbers of assemblages of particles, such as nuclei or atoms.

Of the three mechanisms for generating magnetic fields at the nucleus, the first is by far the most important in the case of iron, while the second mechanism is frequently dominant in the case of rare-earth metals.

A paramagnetic substance differs from a magnetically ordered one in that, although the atoms have magnetic moments, these are not coupled together, but fluctuate in a random way under the influence of thermal agitation. Fields are produced at the nuclei in exactly the same way as before; but they fluctuate so rapidly that the nuclei cannot follow them; the nuclei 'sense' a field of zero. If an external magnetic field is applied, it becomes energetically favourable for the atomic moments to align with the field, but this energy gain has to compete with the loss of entropy that would be associated with the loss of random movement of the atomic moments. The nett result is that although all the moments continue to oscillate rapidly, there is a tendency to align with the magnetic field. The 'thermally averaged' field at the nucleus is no longer zero, the nucleus senses this, and the Mössbauer spectrum now provides a measure of the degree of magnetic alignment of the sample.

Combined Quadrupole and Magnetic Interactions

If a nucleus having a spin greater than one is subjected to a magnetic field and an efg acting along the same axis the problem is easily solved, since the energy perturbations are additive. If we return to Fig. 5.2b and add a magnetic perturbuation, we see that in both the ground and excited states, the pairs of substates will split, with the + 3/2 substates in the excited state being affected most. The + 1/2 to + 1/2 transition will split up into four separate lines. For the + 3/2 to + 1/2 case two of the transitions are forbidden by the $\Delta_I = 0, \pm 1$ selection rule, and so this transition splits only into two. This experiment therefore determines the sign of the efg. If the magnetic field and the efg are not acting along the same axis the problem is not so easy, as it is no longer clear which axis should be taken as the quantisation axis. (The problem can be solved readily on a computer). However the basic result remains, - the + 1/2 to + 1/2 transition splits into four and the + 3/2 to + 1/2 transition splits into two. This ensures that for any sample it is always possible to determine the sign of an efg by applying a strong enough magnetic field.

Note on Quantisation

Quantisation is a necessary consequence of the wave-like
properties of matter. Given that nuclei and atoms, and their
constituent particles, protons, neutrons, and electrons, have
to be treated in terms of wave mechanics, we will want to be
able to write down a function, known as the wave-function,
from which we can derive all the observable properties
required (sometimes we are able to do this exactly; in most
cases we have to be content with approximations). Now there
are several requirements that an acceptable wave-function must
satisfy. For a particle in a stationary state, that is one in
which it is not exchanging energy with its environment, two
criteria are that the wave-function must be single-valued
(the particle must not have more than one value of a given
property at any point in space) and that it must be continuous
(we do not expect the properties of the particle to change
abruptly). Consider the case, which occurs frequently in this
book, of a particle situated in an environment having rotational symmetry about a given axis. In order to satisfy the
above criteria the wave-function must be such that any
circular path around the symmetry axis passes through an
integral number of wave-patterns. Or, more easily visualised,
the wave pattern about the axis must be characterised by a set
of planar modes, with the symmetry axis as a common line of
intersection. Now the angular momentum of the particle about
the axis is proportional to the number of such nodes, and we
see that the angular momentum is quantised: it may only
assume values which are integer or half-integer multiples of
$h/2\pi$, where h is Planck's constant. (Actually the states for
which the angular momentum is a half-integer multiple of $h/2\pi$
are only invariant with respect to rotation by 4π about the
symmetry axis, rather than 2π, but, for somewhat esoteric
reasons, this is acceptable.)

A nucleus may contain between one and several hundred
nucleons — protons and neutrons — each with an associated
angular momentum. In a given nuclear state these momenta will
add vectorially, many of them cancelling out, to give the nett
observable angular momentum, $n(h/2\pi)/2$, where n is an integer.
The integer or half-integer, $n/2$, is known as the nuclear spin
quantum number, I. The nett angular momentum will generate a
nuclear magnetic moment, μ_N. If I is greater than $1/2$ the
nucleus may also simultaneously have a quadrupole moment,
colinear with the magnetic moment. If we put such a nucleus
in a magnetc field, its energy will be dependent upon its

orientation with respect to the field. However, the field defines a symmetry axis, about which the angular momentum must be quantised. We find that the magnetic moment may only assume a certain set of orientations with respect to the field; they are those for which the components of the angular momentum along the field direction are $Ih/2\pi$, $(I-1)h/2\pi$, $(I-2)h/2\pi$, ... $-Ih/2\pi$. The same set of restrictions applies to a nucleus with a quadrupole moment subjected to an electric field gradient - the allowed orientation with respect to the gradient are governed by the angular momentum quantisation rules. The set of numbers, $I,(I-1),(I-2)$ etc. are given the label M_I where the subscript indicates that it is nuclear states that are being referred to.

Let us consider the simplest case that will illustrate the point, a nucleus having $I=1$ and a positive quadupole moment. The allowed values for the angular momentum around a given axis are $h/2\pi$, 0, $-h/2\pi$. If a magnetic fild is switched on the first of these is raised in energy, the last is lowered, and the remaining substate is unperturbed. If a positive electric field gradient is switched on, the first and last states go to higher energy (remember that the quadrupole interaction is a function of $\cos^2 \theta$, where θ is the angle between the quadrupole moment and the electric field gradient) and the remaining substate goes to lower energy.

REFERENCES

Kuhn, H.G. Atomic Spectra, Longmans, Green and Co. Ltd., 2nd Ed. (1969).

CHAPTER VI

APPLICATIONS TO CHEMISTRY

The solid state chemist is interested in the chemical states and electronic configurations of atoms in solids, the spatial arrangements of these atoms, and the ways in which they interact with one another over both short and long distances. In suitable cases Mössbauer spectroscopy can provide valuable information on all these aspects. Such information, while often unique, is rarely complete, and the Mössbauer method must be augmented with other techniques. This point is illustrated very well by the case of one of the simplest of coordination compounds, potassium ferricyanide, $K_3Fe(CN)_6$. Here it has not yet proved possible to find a theoretical model which accounts simultaneously for the magnetic susceptibility results, the electron spin resonance results and the Mössbauer results. It is clear that each technique is reflecting a different aspect of the true molecular wave-function and until we can construct accurate wave-function ourselves we will need results from a range of techniques before we can be sure of obtaining a true picture of any given example.

In an earlier chapter it was seen that there are three principle parameters that may be obtained from a Mössbauer spectrum – the isomer shift, the quadrupole splitting and, in the case of magnetic compounds, the magnetic hyperfine splitting. The chemical significance of each measured parameter is often treated separately, but it is a common experience that while a single parameter may be ambiguous in isolation, a group of parameters, measured simultaneously, is much more revealing. This chapter has therefore been arranged in terms of the problems that may be tackled rather than in terms of the information that may be gained from specific hyperfine parameters.

Chemical States and Electronic Configurations

In as much as the different states available to an atom may have different numbers of s electrons we would expect them to have

different isomer shifts. Thus we find that there is a large
change in isomer shift on passing from, say, Sn(II) $(5s)^2$ to
Sn(IV) $(5s)^0$ or from Sb(II) $(5s)^2$ to Sb(V) $(5s)^0$. However
this is not the whole story. The Sn(II) ion, for example, does
not usually exist in isolation; it will probably be bonded
covalently to nearby atoms and this invariably involves some
mixing of asymmetric 5p wave-functions of the tin atom into
the spherical 5s functions. This, by reducing the symmetry,
produces an electric field gradient at the tin nucleus and a
quadrupole splitting in the Mössbauer spectrum; and since the
5s functions are now inter-mixed with 5p the isomer shift is
also affected. So in order to understand the isomer shifts
of Sn(II) compounds we have also to study the quadrupole
splitting, and vice versa. We find in general that Sn(II) is
characterised by positive isomer shifts and large quadrupole
splitting compared to Sn(IV) compounds. Similar rules can be
found for other elements.

There is a second rather more subtle way in which the isomer
shift is dependent upon the state of the atom. The wave-
function of an s-electron is not independent of the number and
type of other electrons present – this is the so-called
screening or shielding effect. We find that on passing from
Fe(II) $(3d)^6$ to Fe(III) $(3d)^5$ the s-electron density at the
nucleus has increased, not because there are more s-electrons,
but because a $(3d)^5$ shell is less effective at screening the
nuclear charge than a $(3d)^6$ shell, and has allowed the s-electrons (mainly 3s) to crowd in closer to the nucleus.
Similarly large differences are observed between the isomer
shifts of Eu(II) compounds and Eu(III) compounds, and between
Au(I) compounds and Au(III) compounds.

The chemical state of an atom is not necessarily defined
completely by its valency: we may be uncertain about the
electron configuration. This is relevant in transition metal
compounds where the metal ion may be in the high-spin state,
with the maximum number of unpaired valence-shell electrons,
in the low-spin state, with the minimum number, or occasionally
in an intermediate position. The Mössbauer parameters frequently allow us to distinguish between these possibilities. The
mechanisms by which mere changes of configurations bring about
changes in isomer shift are not understood completely, although
the rearrangement of electrons which takes place on passing
from the high-spin to the low-spin state is always one which
allows the atom in question to come into closer contact with
its neighbours, resulting in increased covalency. As we shall

see later this invariably involves changes in the s-electron population on the metal. The quadrupole interaction will also be affected: in the case of Fe(II) the change from high to low spin involves changing from the asymmetrical configuration $(t)^4 (e)^{2*}$ to the spherically symmetric $(t)^6$ configuration and therefore brings about a large decrease in the quadrupole interaction. So high-spin ferrous compounds are characterised by large quadrupole splitting and positive isomer shifts, compared to low spin compounds.

The corresponding distinctions in the case of ferric compounds are, unfortunately, nothing like so clear cut, although the observed trends are more or less as expected. In the few cases where distinction cannot be made on the basis of isomer shift and quadrupole data, measurements of magnetic hyperfine effects usually resolve the difficulty.

With these considerations as a basis for interpretation, Mössbauer spectroscopy plays a useful role on the purely analytical level. It is particularly helpful when a Mössbauer isotope is present in several different chemical forms within the same sample - a common situation in the study of minerals, ores or corrosion products; not only can the various species be identified, but the relative amounts can be determined.

Structural Considerations

There is no systematic way that Mössbauer spectroscopy can be applied to the determination of an unknown crystal or molecular structure. However a Mössbauer spectrum frequently provides helpful pointers, or is instrumental in eliminating postulated

---

*The five d orbitals are degenerate (all of equal energy) in the gas phase. On entering an octahedral site, (the most common and illustrative case) two of the orbitals are destabilized with respect to the other three by virtue of the fact that they are forced into closer proximity to the negative changes in the crystal. The upper pair of orbitals are labelled 'e', the lower triplet 't'. In most cases the octahedron is distorted, causing further splittings within both the e and the t subsets. Technically the levels should now be re-labelled, but it is simple and convenient to retain the labels e and t, since they still describe the broad features of the situation.

structures. One of the early successes in this field was the discovery that one of the three iron atoms in the compound $Fe_3(CO)_{12}$ was significantly different from the other two, eliminating all postulated structures in which the iron atoms were identical. A more recent example made use of a Mössbauer spectroscopy as a finger-printing technique. The compound $[ReCl(N_2)(Ph_2PCH_2CH_2PPh_2)_2]^+(FeCl_4)^-$,* of interest in the study of nitrogen fixation, and at that time of unknown structure, was shown to contain the $(FeCl_4)^-$ group, since the distinctive features of the $(FeCl_4)^-$ spectrum were evident in the Mössbauer spectrum of the compound.

Many of the applications of Mössbauer spectroscopy in structural work are concerned with the question of symmetry. Mössbauer spectroscopy has proved itself curiously adept at detecting small deviations from cubic symmetry or from rotational symmetry about the Mössbauer site. These successes have often been due to the multiplying effect of an open shell configuration, but have been useful nevertheless. Consider the high spin ferrous configuration appropriate to octahedral coordination (t^4) (e^2). It is helpful to consider this as a single t election operating outside the spherically symmetric half-filled shell $(t)^3$ $(e)^2$. Placed in a cubic environment this configuration would generate no field gradient at its centre because the t electron would time-average itself equally between the three available t orbitals and thus retain spherical symmetry. However only a very small distortion is required to split the t subset by energies greater than kT, and in this situation the electron will occupy the lowest energy orbital preferentially. In doing so it may generate an electric field gradient at the centre which is many times greater than the field gradient that would have been generated by the original distortion. It was through this phenomenon that the ion $(FeCl_4)^{--}$ was first found to be distorted from tetrahedral symmetry.

Although the multiplying effect of an open shell produces large effects which are easy to measure, the detailed interpretation is not easy. From the structural point of view closed shell systems are often easier to understand. One of the successes in this field has been in the ability of Mössbauer spectroscopy to distinguish between cis/trans isomers. If M is a closed shell metal atom the electric field gradient generated at the M nucleus in the octahedral complex

*Ph = C_6H_5

trans MA_4B_2* is twice as great as, but of opposite sign, to
the electric field gradient generated at the M nucleus of the
cis-isomer. (This follows from simple point charge considerations and is an example of the success of this simple concept).
This useful result is widely applicable in compounds of low-
spin Fe(II) and of Sn(IV).

Interatomic Interactions

For the chemist the most important interaction between atoms
is that which results in delocalization of the electronic
wave-functions over two or more atomic centres, the extreme
form of which is the covalent bond. There has been no
shortage of Mössbauer studies in this field, and they break
down basically into two types – studies of fairly complex
systems in which the aim is to learn something about the
bonding from the Mössbauer parameters, and studies of simple
systems where the aim is to establish precisely which properties of molecular wave functions the Mössbauer parameters are
actually measuring. Ideally the second kind of study would
have predated the first, but in practice has proved extremely
intractable. None of the three main hyperfine parameters is
susceptible to straightforward interpretation, and it is
important and instructive to see why this is so. If the
isomer shift had been a direct measure of the s-electron
population chemists would have had a powerful tool for studying the specific involvement of s-electrons in covalency.
This would have had a profoundly simplifying influence. However, the realization that screening of s-electrons by other
electrons was of an importance comparable to that of the
s-electron population added an unwelcome complication.

The interpretation of quadrupole interactions is equally
difficult, mainly due to what is known as the Sternheimer
anti-shielding effect. If a negative charge is brought up to
a spherically symmetric closed shell atom the outer regions
of the electron cloud are driven from that side of the atom
to the other, and also towards the centre, as indicated in
Fig. 6.1. The net result is a pile-up of negative charge
along the perturbation axis near the nucleus, greatly

---

*In the trans octahedral complex MA_4B_2 the B atoms are oppo-
 site one another across the octahedron: in the cis isomer
they occupy adjacent positions on the octahedron.

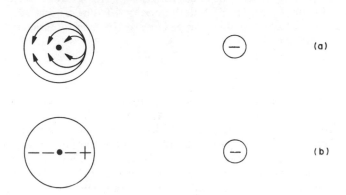

Fig. 6.1   The effect of a perturbing negative charge on the
electron distribution of a nearby atom.   The electron move-
ments are indicated in a, and in b the final charge distribution is shown, where the small changes indicate changes from
the original spherical symmetry.

enhancing the quadrupole coupling.   Multiplication factors in
excess of 10 are quite common.

The isomer shift and the quadrupole coupling are very similar
in that a detailed understanding can only be contemplated in
terms of wave-mechanical calculations involving all the
electrons in a molecule.   Such calculations are feasible only
for small molecules, and even then are lengthy and difficult.
In the past few years however, Mössbauer parameters for simple
molecules such as Fe_2, FeO, $FeCl_2$, Sn_2, SnO, SnO_2, I_2, etc.
have become available, and initial attempts at interpreting
these parameters are encouraging.   (The measurements are made
with the molecules isolated in inert matrices such as solid
argon, ensuring that the Mössbauer parameters are reflective,
as nearly as possible, of purely diatomic molecular wave-
functions).

The interpretation of magnetic hyperfine interactions is
complicated by the fact that, as we indicated in Chapter 5

there are often three separate contributions. If single
crystal specimens are available it is possible to distinguish
between the Fermi contact term, the orbital term and the
dipolar term, due to the anisotropic nature of the latter two.
However, good single crystal specimens of compounds of interest
are frequently unavailable. In cases where the orbital and
dipolar terms are absent (for example in compounds of the
high-spin Fe^{+++} ion, which has no orbital momentum and no net
dipolar term) there is considerable debate regarding the
effect of covalency on the Fermi contact interaction.

It would, however, be totally wrong to conclude from the above
that Mössbauer parameters have no relevance in studies of
covalency. If it is accepted that the multiplicative factors
such as shielding and anti-shielding effects are going to be
similar in similar cases, a great deal can be learned from
studies of groups and series of compounds. Consider a low-
spin FeII atom with the 3d electronic configuration $(t)^6(e)^0$
in the gas phase. The 3d and $3d_{x^2-y^2}$ orbitals are empty, as
are the 4s orbitals. Bring up to it, say, a CO molecule, to
form a linear complex, Fe-CO, with Z as the axis of symmetry.
In the complex formation some of the electrons from the $2p_z$
and 2s orbitals of the carbon atom will be partially 'donated'
into the 4s orbitals of the Fe atom (causing the isomer shift
to become more <u>negative</u>) and to the $3d_{z^2}$ orbitals (producing a
<u>negative</u> efg). But suppose that the $(t)^6$ orbitals become
involved, and donate charge back into the partially empty $2p_x$
and $2p_y$ orbitals of the carbon atoms. The shielding of the
s electrons on the iron atom will decrease, increasing the
electron density at the nucleus, again driving the isomer shift
<u>negative</u>. But by removing charge from near the z axis the
back-donation will make a <u>positive</u> contribution to the efg.
So if we looked at a series of compounds FeX where X is
variable, and found a spread of values, but a positive corre-
lation between the isomer shift and the efg, we would be
entitled to conclude that simple donation of electrons into
the 4s and $3d_{z^2}$ orbitals of the iron atom was the predominant
bonding mechanism, and that the propensity to donate was
different for different X's. However, if we found a few
'rogues' with unexpectedly positive efg's we would have grounds
for thinking that some back-donation of t electrons was
occurring in such cases. It is on this sort of level that the
Mössbauer effect is very useful in studies of chemical bonding,
and the type of argument employed here has been successfully
applied to the bonding behaviour of many closed shell atoms.

The magnetic hyperfine interaction can also be employed in a diagnostic way. We saw in Chapter 5 that in paramagnetic samples subjected to an applied magnetic field, the field experienced by the nucleus was dependent upon the magnetisation of the samples. Now the magnetisation of the sample is frequently anisotropic, reflecting considerable subtleties in the way chemical bonding effects the energy levels of atoms. We also saw in Chapter 5 that the type of magnetic splitting we get in Mössbauer spectra are dependent upon the angle between the magnetic field and the principal efg. The upshot is that from a powdered specimen it is possible to determine, for example, the temperature dependence of the three independent components of the magnetisation, provided that there is an efg present. Admittedly such data could be obtained from magnetic susceptibility measurements, providing that good single crystals were available, and providing that in the crystal structure the molecules were not paired in inconvenient ways (for example, with their respective z axes perpendicular). Or electron paramagnetic resonance might prove helpful, provided that the lowest-lying electronic spin-state of the molecule is degenerate, or nearly so, that the substance can be prepared in a magnetically dilute form, and that there are no insuperable problems with relaxation phenomena. Such conditions are frequently not fulfilled, and Mössbauer spectroscopy may then be the only method available for studying the magnetic anisotropy in detail.

In conclusion we can say that Mössbauer spectroscopy is a useful addition to the investigative techniques available to the chemist. Like any other technique, it has its drawbacks and limitations, but it often produces results when all other methods fail.

For more detailed accounts of this large field the reader is referred to the books by N.N. Greenwood and T.C. Gibb, ('Mössbauer Spectroscopy', published by Chapman and Hall, 1971), V.I. Goldanskii and R.H. Herber, ('Chemical Applications of Mössbauer Spectroscopy', published by The Academic Press, 1968), and by G.M. Bancroft ('Mössbauer Spectroscopy', published by the Halsted Press, 1973).

CHAPTER VII

APPLICATIONS TO METALLURGY

The Mössbauer spectrum is a sensitive indicator of the chemical nature of the atom containing the Mössbauer nucleus and its immediate surroundings. In this chapter we give some examples of the way in which information from Mössbauer spectra is helping to solve problems in the study of metals and alloys.

For the first example, we consider the alloy Fe_3Al. X-ray evidence shows that this substance can exist in an ordered form if cooled slowly from a high temperature or in a disordered form if quenched rapidly. The ordered structure is shown in Fig. 7.1, where it can be seen that the iron atoms lie on two inequivalent sites, labelled A and D sites. The A sites have four Al atoms and four Fe atoms in their first neighbour shell, and D sites have all eight first neighbour positions occurpied by iron atoms. For each D iron site, there are clearly two A sites. The Mössbauer spectra of the order and disordered forms are shown in Fig. 7.2. It is clear that the spectrum of the ordered substance is made up mainly of two 6 line spectra of relative intensities 2 and 1, with different values of the hyperfine field and isomer shift. These two spectra can be identified with the A and D site iron atoms. A small third component is assumed to be due to the occasional occupation of an Al site by Fe. The spectrum of the disordered substance on the other hand shows only one set of 6 broad lines. This corresponds to the fact that iron atoms can now be found in a wide range of environments amounting almost to a continuous distribution.

In this case, the Mössbauer spectra have added little to our knowledge, because Fe_3Al had been well studied by X-rays. On the other hand, in the case of Ni_3Fe, the study of ordering by X-rays is almost impossible because X-rays are scattered nearly equally by iron and nickel. However, the narrowing of the lines of the Mössbauer spectrum, and the change of hyperfine field gives a clear indication when ordering occurs.

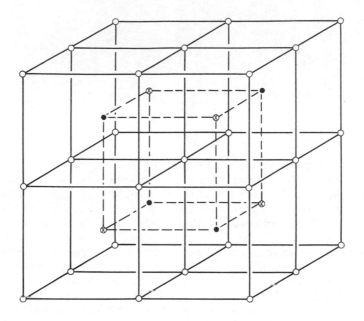

○   Fe on A sites

⊗   Fe on D sites

●   Al on D sites

Fig. 7.1 The lattice of ordered Fe$_3$Al.

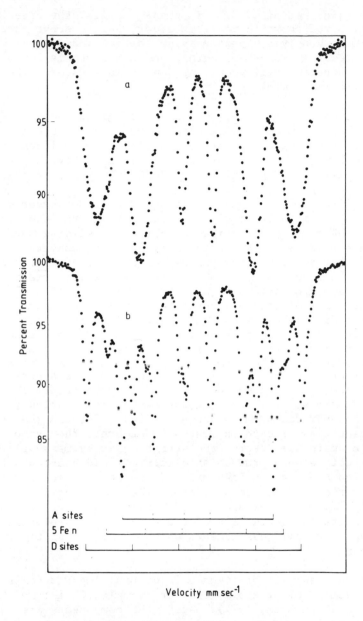

Fig. 7.2a. The Mössbauer spectrum of disordered Fe$_3$Al, (a)
of ordered Fe$_3$Al, (b).

(Drijver, et al 1975). In these studies, the ordering transition was shown to be first order, and to occur partly homogeneously and partly by nucleation and growth. The existence of a two phase region, and a hysteresis zone, apparently of magnetic origin could also be demonstrated. (See also Gros and Pauleve 1970).

In iron-silicon alloys which are of great industrial importance as high permeability materials, Mössbauer spectra have shown that ordering occurs at concentrations of silicon which are too low for X-ray super-lattice lines to have a detectable intensity. (Papadimitriou and Genin, 1972, Cranshaw 1977).

A quite different example of the use of Mössbauer spectroscopy in metallurgical investigation may be found in the study of the changes occurring in lubricated sliding surfaces. In this case, spectra are taken by counting conversion electrons, so that a layer a few hundred Angstroms thick at the surface of the specimen is sampled. The main components found in the surface are martensite, a b.c.t. structure of iron with small additions of carbon, austenite, a f.c.c. structure of iron with dissolved carbon, and Fe_3C, iron carbide or cementite. Martensite is ferromagnetic, and has a Mössbauer spectrum similar to that of pure iron. Austenite is paramagnetic, and its Mössbauer spectrum is a single line. Fe_3C is a complex structure which is antiferromagnetic with a hyperfine field of about 19.0T compared to 33.0T in pure iron. In favourable conditions, all these components can be detected in the Mössbauer spectrum as shown in Fig. 7.3a, and the relative quantities of each estimated. Moreover, the carbon atoms in the austenite destroy the strict cubic symmetry of the f.c.c. structure, and produce an electric field gradient at iron atoms which are their neighbours. These atoms then show a doublet superimposed on the singlet of the austenite line as shown in fig. 7.3b. A measurement of the intensity of the doublet permits the estimation of the amount of dissolved carbon. These are all measurements which are quite difficult by other means.

Another subject of immense industrial importance is the heat treatment of steels. (See for example, Christ and Giles, 1968, Ae and Schwartz, 1974). Again the components are mainly martensite, austenite and carbide, and the intensities of the components in the Mössbauer spectra are nearly proportional to the relative composition of the specimen. The effects of carbonising and nitriding steels have also been extensively

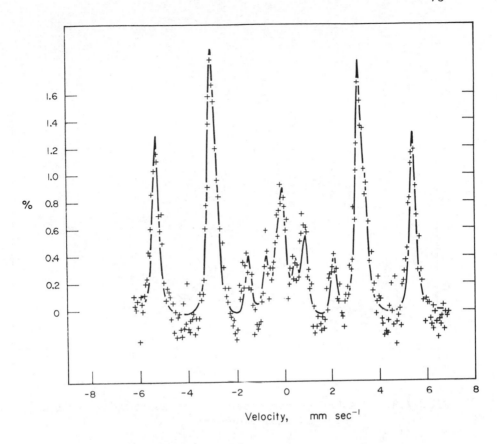

Fig. 7.3a A backscatter spectrum of a steel surface showing
martensitic, austenitic and carbide components.

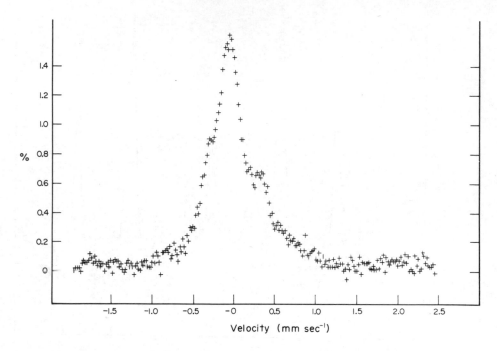

Fig. 7.3 b. A backscatter spectrum of austenite. Note
enlarged scale.

investigated by Mössbauer spectroscopy (Gielen and Kaplow, 1967, Ino et al 1968).

An interesting example of the use of Mössbauer spectroscopy for phase analysis is the study of the solution of iron in copper. (Window, 1972). At room temperature the solid solubility of iron in copper is extremely low. However, if a specimen of copper containing iron in solution at 900°C is rapidly quenched to room temperature, a metastable super-saturated specimen can be obtained, the Mössbauer spectrum of which shows a single line. A 1% alloy will show the single line, and a doublet due to pairs and clusters of small numbers of iron atoms. The pairs and clusters give a doublet spectrum because the iron atoms are not in positions of cubic symmetry, and an electric field gradient exists at the iron nuclei. Moreover, the electron density at the iron atoms in the pairs is different from the solution iron, so that the centre of the doublet does not coincide with the line due to solution iron. If the temperature is raised, diffusion of the iron atoms occurs so that the number of isolated atoms falls, and the number of pairs or clusters increases.

When diffusion is allowed to proceed further, precipitates of f.c.c. iron, called γ-iron, form coherently with the copper lattice. γ iron is the stable form of iron above 910°C. The γ-iron precipitates are found to be antiferromagnetic with a Neel temperature of about 20K and a hyperfine field of about 2T. On plastic deformation of the sample, the γ-iron precipitates transform into the normal α-iron phase, so that the familiar 6 line spectrum of α-iron appears.

REFERENCES

Abe, N and Schwartz, L.H. (1976). Mat. Sci. and Eng. 14 239.
Christ, B.W. and Giles, P.M. (1968). Trans. Met. Soc. AIME 242 1915.
Cranshaw, T.E. (1977). Physica 86 88B 391.
Drijver, J.W., van der Woude, F. and Radelaar, S. (1975). Phys. Rev. Lett. 34 1026.
Gielen, P.M. and Kaplow, R. (1967). Acta. Met. 15 49.
Ino, H., Moriya, T., Fujita, F.E., Maeda, Y., Ono, Y. and Inokuti, Y. (1968). J. Phys. Soc. Japan, 25 88.
Papadimitriou, G. and Genin, J.M. (1972). Phys. Stat. Sol.(a) 9 K19.
Schwartz, L.H. and Kim, K.J. (1976). Metall. Trans. A 7A 10 1567.

Window, B. (1972). Phil. Mag. 26 681.

CHAPTER VIII

APPLICATIONS TO MAGNETISM

Ferromagnets

One of the outstanding problems in solid state physics is the
mechanism of the spontaneous magnetization of iron, cobalt and
nickel. The dependence of the magnetization on temperature is
quite well represented by the molecular field approach, in
which each atom is imagined as acted upon by a field produced
by all the other atoms. To account for the observed values
of the Curie temperature, the field has to have a strength of
about 5.10^6 Oe, several orders of magnitude larger than the
field due to the magnetic moments of the other atoms.

The coupling between magnetic atoms was shown by Heisenberg
to lie in the "exchange interaction", which is mentioned
briefly in "Notes on Quantization" in Chapter V. We are there
considering electrons on the same atom, the so-called intra-
atomic exchange, but the same argument holds for electrons on
different atoms, or for electrons on atoms and itinerant
electrons in a metal, provided that their wave functions overlap. The exchange interactions align the moments in a metal
and produce the spontaneous magnetization. The ease with
which hyperfine fields can be measured by Mössbauer spectro
scopy leads one to hope that it might be very helpful in the
study of magnetism.

This hope for quantitative information has been largely frus-
trated by the great complexity of the mechanisms producing
hyperfine fields. As a first approximation it was assumed
that the hyperfine field was nearly porportional to the moment
on the atom. In some cases this appeared to be true. For
example, the temperature dependence of the hyperfine field at
^{57}Fe nuclei in iron is close to the temperature dependence of
the magnetization. (Fig. 8.1) In ordered Fc_3Al, (see Chapter
VII) at the two iron sites the hyperfine fields are found to
be roughly proportional to the magnetic moments which have

78

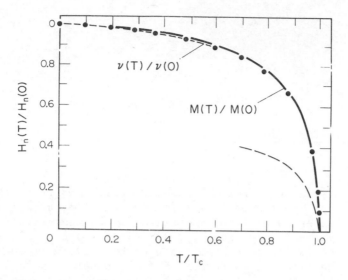

Fig. 8.1 The reduced hyperfine field (ν) and magnetization (M) as a function of temperature, (Hanna et al 1962) been measured by neutron scattering.

That this approach is too simple was realized when Samoilov (1959) showed that a large hyperfine field existed at Au nuclei in iron. Gold is a diamagnetic atom, and is most unlikely to have any moment on it. We clearly cannot neglect the fact that a component of hyperfine field can be produced by its neighbouring atoms. For example, such a field may be produced by conduction electrons polarized by the other atoms and having a finite density at the nucleus under consideration. They then produce an effective field by the Fermi contact interaction. Other atoms may also polarize s-electrons in closed shells by exchange interactions.

Gold, tin, and antimony are all Mössbauer elements in which spectra can be obtained fairly easily, and a large area of investigation has now opened up measuring the hyperfine fields at these elements in a variety of magnetic environments. The hyperfine fields at many nuclei can be measured by other methods (e.g. perturbed angular correlations, NMR) and attempts can now be made to explain the systematic dependence of hyperfine field on charge, or period number in the periodic table.

Evidence has been accumulated to show that there are many interactions which can produce large values of effective field at the nucleus, and that these contributions can be of either sign. The observed field is therefore the relatively small resultant of many contributions, any one of which may dominate in the appropriate circumstances. An interesting case is that of ^{119}Sn in iron and cobalt. Here there appear to be two main contributions to the field which nearly cancel. The temperature dependences of the two components are presumed to depart slightly from the temperature dependence of the magnetization, leaving a small resultant which departs very considerably. In cobalt, the tin hyperfine field is believed even to change sign. (Fig. 8.2).

Even though an exact relationship between hyperfine field and local magnetic moment has not been found, Mössbauer spectroscopy is a sensitive indication of magnetic properties. The variation of the mean hyperfine field in an iron alloy is qualitatively similar to that of the average atomic magnetic moment of the alloys as a function of composition (the Slater-Pauling curve).

Mössbauer spectroscopy can be used to study the disturbance produced in an iron lattice by the introduction of an alloying element. A spectrum of a dilute alloy usually shows a component closely resembling the spectrum of pure iron, with additional components with smaller fields (larger in the case of Co and Ni) and a slightly changed isomer shift, believed to be due to those iron atoms which have one or more atoms of the alloying element in its first neighbour shell. By using single crystal specimens and varying the magnetization axis, it is possible to measure the electric field gradient and dipolar magnetic field at iron atoms which are first or second neighbour to a solute atom. It is found that for the elements Cu to As, the magnitude of the disturbance is a linear function of Z. (Cranshaw, 1977).

The mechanism of the disturbance is usually sought in the "screening" of the charge on the solute atom. When a foreign atom is dissolved in the iron, we have on one lattice point an element whose charge is different from that on the other lattice points. If the difference is positive, there will be an attractive potential, and itinerant electrons will spend more time in its vicinity, thus reducing or "screening" the charge. If the difference is negative, the opposite occurs. One might expect that this extra density of charge would decrease smoothly with distance from the solute. This however

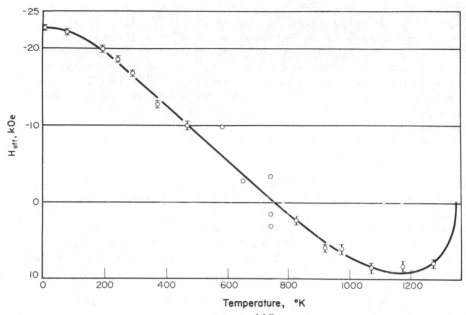

Fig. 8.2 The hyperfine field at ^{119}Sn in FCC cobalt
as a function of temperature.

is not the case, because the upper limit of the energy of the
electrons is the Fermi energy, and the lower limit of wavelength the Fermi wavelength, $1/k_F$. This turns out to have a
value of the same order as the ionic diameter, and therefore
we have a typical diffraction problem. Round the solute there
exist haloes of charge density whose separation is approximately the Fermi wavelength (Friedel oscillations, Friedel,
1958). In magnetic iron, the diffraction or scattering of
the electrons may clearly be spin dependent. In that case, as
well as haloes of charge density, there will be haloes of
spin density, and these will show up as changes of the hyperfine field at neighbouring atoms.

Antiferromagnets

The Mössbauer Effect has also made important contributions to
the study of the antiferromagnetic crystals. In antiferro-
magnets the exchange produces an antiparallel rather than a
parallel alignment of the spins. In the simplest cases (e.g.
FeF_2) there are two-sublattices, with magnetizations of
opposite signs. Thus below the transition temperature T_N

(the Néel temperature) there is spontaneous magnetic order,
but no macroscopic moment as in ferromagnets. Hence conventional magnetization measurements are not very effective in
providing data on these crystals. However, a hyperfine field
is observed in the Mössbauer spectra below T_N, and assuming
that it is proportional to the sub-lattice magnetization, the
latter can be measured as a function of temperature.

The decrease M_0-M in sub-lattice magnetization near OK from the
saturation value M_0 may be observed to test the T^3 law of spin
wave theory. Just below T_N M is proportional to $(T_N - T)^\beta$,
where the exponent β has been found to be about 1/3, in agreement with the theory of critical phenomena.

By making measurements of the hyperfine spectrum of a single
crystal absorber of an antiferromagnet, the axis of spin
alignment may be determined. $\Delta m_I = 0$ transitions have zero
intensity when the γ-rays are parallel to the magnetic moments,
and a maximum intensity when the γ-rays are perpendicular to
them.

Paramagnets

In a typical paramagnet the hyperfine field is averaged to zero
by the rapid fluctuations of the atomic magnetic moments.
However it may be possible to observe magnetic hyperfine
splitting if:

(a) the fluctuations are slow i.e. if the electron spin-spin
     and spin-lattice relaxation times are large, or if

(b) an external magnetic field large enough (e.g. several
     Tesla at 4.2K) to produce an appreciable magnetization is
     applied.

Examples of the former type are found in Chapter X on biological molecules. Since the iron atoms in proteins are well
separated from each other by the amino-acid chains the spin-
spin relaxation is slow, and at low temperatures (where spin-
lattice relaxation is also slow) hyperfine splitting is
observed.

The application of an external field B to a crystal containing
atoms with magnetic moment μ produces a hyperfine splitting
which may be described as an internal field B_{int} proportional
to the magnetization produced:

$$B_{int} = \frac{M \frac{\mu B}{nT}}{M_0} B_n{}^{(0)}$$

where $B_n{}^{(0)}$ is the saturated value of the field at the nuclei. For small values of the magnetization

$$B_{int} = \frac{\chi \frac{B}{\mu_0 N \mu}}{} B_n{}^{(0)}$$

where χ is the molar susceptibility. Thus measurements of the hyperfine splitting may be used to deduce the magnetic susceptibility of the atoms. For ferric ions the susceptibility is almost isotropic, and may be used to determine the zero field splitting of the ground states by the crystal field. For ferrous ions the susceptibility is anisotropic, and from measurements on single crystals with the external field applied along different axes the contributions to the hyperfine field from the spin and orbital motion of the 3d electrons may be distinguished.

In the metallic state the nature of the magnetic moments on the atoms is of special interest, and gives rise to questions, such as how localized is the moment and under what conditions does an iron atom have a magnetic moment?

An important measurement has been carried out on CuFe alloys by Frankel et al (1967) using the Mössbauer Effect. For alloys very dilute in iron the saturated hyperfine field B_n observed for large values of B/T was found to depend upon the value of the applied field B at low temperatures. The larger B was, the larger the observed value of B_n. This provides direct evidence for the Kondo Effect – the destruction of the localized magnetic moment at low temperatures by antiferro-magnetic coupling between the d-electrons on the iron and the conduction s-electrons. Above a temperature T_K of about 16K the alloy behaves like a normal paramagnet. At lower temperatures the magnetic moment is completely compensated, but can be partially restored by the application of the field which effectively decouples the spins of the d- and s-electrons.

When molten mixtures of certain metal-metalloid combinations with the approximate composition $M_{80}A_{20}$ (M = Fe, Co, Cu, Pd and A = P, C, B, Si) are quenched at rates of about 10^6 degrees sec^{-1}, the resulting material may have no crystalline structure, but is amorphous. These substances sometimes have favourable magnetic and physical properties, and are of great industrial importance.

The Mössbauer spectra of these alloys show broad lines, indicating a distribution of hyperfine fields, as would be expected. Various models have been used to fit the spectra, for example, dense random packing of spheres, or very small crystallites, and it is not yet clear which is the most appropriate. On the other hand Mössbauer spectroscopy can certainly reveal the temperature dependence of the components, and enables the process of crystallization to be studied.

REFERENCES

Cranshaw, T.E. (1977). Physica 86-88B 443.
Frankel, R.B., Blum, N.A., Schwartz, B.B. and Duk Joo Kim. (1967). Phys. Rev. Letters 24 1051.
Friedel, J. (1958). Nuovo Cimento Supp. VII 287.
Hanna, S.S., Preston, R.S. and Heberle, J. (1962) Phys. Rev. 128 2207.
Samoilov, B.N., Sklyarevski, V.V. and Stepanov, E.P. (1959). Sov. Phys. J.E.T.P. 36 448.

CHAPTER IX

RADIATION DAMAGE

The contribution of Mössbauer spectroscopy, or indeed of any microscopic technique to general studies of radiation damage has not so far been very great. Even in heavily irradiated material only about one lattice site in 10^5 is actually 'damaged', and probe atoms within the lattice will have only a small probability of being close to such a site. The situation is made worse by the tendency of damaged sites to migrate and then congregate in clusters of various kinds. However there are certain aspects of radiation damage which Mössbauer spectroscopy is uniquely suited to study. Mössbauer emissions usually occur at the end of a more or less energetic radioactive decay chain. The energy liberated during the decay will have been expended in damaging the nearby lattice and the Mössbauer atom, having caused the damage, is 'on the spot' to record the effects of it.

As a simple example, consider the case of an isotope which decays to an upper Mössbauer level by means of a nuclear event liberating a few tens of eV available energy. This energy may be sufficient to eject the atom from its lattice site, displacing one of its neighbours and leaving behind a vacancy. The displaced neighbour will displace another, and so on in a shunting motion. Finally as energy is dissipated at each displacement a somewhat distant atom will get pushed into an interstitial site. The net result will be that the Mössbauer atom is situated in a lattice that is normal apart from the vacancy next to it and an atom in an interstitial position some distance away. When the Mössbauer atom emits its γ-ray the vacancy would be expected to influence the hyperfine parameters, the interstitial probably would not. These ideas seem to be borne out by experiments. In studies by Vogl et al. (1973) dilute solutions of ^{192}Os and ^{196}Pt in α-iron were subjected to neutron irradiation at low temperatures (4.2K). The irradiation produces the isotopes ^{193}Ir and ^{197}Au respectively by neutron capture, followed by β-decay with the release

of 30 eV of recoil energy immediately after the neutron
capture. The samples were removed from the reactor, without
allowing their temperatures to rise above 4.2K, and were used
as sources in Mössbauer measurements. In the case of the ^{193}Ir
resonance, a considerable fraction of the iridium atoms had
hyperfine fields 6% less than that observed for ^{193}Ir in a
normal iron lattice. This is the sort of change in hyperfine
field that would be expected to be caused by an impurity atom
at the nearest neighbour position to a Mössbauer emitter
indicating that in terms of the disturbance it causes, a
vacancy behaves rather like an impurity atom. No quadrupole
interaction was observed at the Mössbauer site and this is
taken to confirm that the vacancy is at the nearest neighbour
position. (In α iron the line joining nearest neighbours is
at 55° to the magnetisation directions which in view of the
strong magnetic interaction is also the quantisation axis.
Thus $3\cos^2\theta-1$ is zero and no quadrupole interaction would be
expected). The fact that a large proportion of the iridium
atoms showed hyperfine parameters indicative of normal iron
lattice is seen as a reflection of the fact that producing a
displacement 'shunt' is easier along some directions than
along others and given the limited amount of energy available
some of the iridium atoms may not have had sufficient energy
to cause displacement in certain directions.

In the case of highly energetic nuclear events preceding
Mössbauer emission the picture is not so favourable. Here the
recoiling atom may tear through the lattice for long distances,
finally displacing an atom from a volume of perfect lattice.
The shunt thus produced will yield a distant interstitial and
the net result is that the Mössbauer emitter is sitting in
'perfect lattice' having caused a distant interstitial and a
more distant vacancy, neither of which will affect the hyperfine parameters. However there is still some information to be
gained in this type of situation. In its progress through the
lattice the recoiling atom will produce a large amount of
local heating, and it is of interest to know how quickly this
energy gets dissipated. We can make use here of the Debye-
Waller factor and reason that a Mössbauer type emission will
occur only if the surrounding lattice has achieved thermal
equilibrium between the time the Mössbauer atom came to rest
and the time the γ-ray is emitted. By using a series of
Mössbauer isotopes of different half-lives it is possible to
estimate the rate at which local thermal equilibrium is reached.
This has been done for a series of rare-earth isotopes, which
have the advantage of being chemically very similar to each

other. The dissipation of the 'heat-spike' is found to occur over a time-scale of about 10^{-11} secs. (Walker and Chien, 1974).

We saw earlier that a Mössbauer isotope dispersed in a lattice has little chance of detecting generalized damage caused by irradiation. However, there are instances where it has proved possible to 'collect' the damage. If, for example, aluminium containing a small amount of ^{57}Co as Mössbauer probe is irradiated with neutrons at low temperature, damage in the form of vacancies and their attendant interstitials, is created. Upon subjecting the lattice to mild annealing the interstitials migrate and attach themselves to the ^{57}Co impurities. Their presence shows up as a quadrupole inter-action at the ^{57}Co sites in the Mössbauer emission spectrum. Further annealing causes the annihilation of the interstitials and the Mössbauer spectrum reverts to that of ^{57}Co in alumin-ium. (Vogl et al. 1974).

Careful studies of the temperature dependence of the Mössbauer spectrum allows the damage annealing process to be followed in great detail.

REFERENCES

Vogl, G., Schaefer, A., Mansel, W., Prechtel, J. and
   Vogl, W. (1973). Phys.Stat.Sol. (b) 59 107.
Vogl, G., Mansel, W. and Vogl, W. (1974). J.Phys.F:
   Metal Phys. 4 2321.
Walker, J.C. and Chien, C.L. (1973). 5th Intnl.Conference on
   Mössbauer Spectroscopy, Bratislava Pt 3, 560.

CHAPTER X

APPLICATIONS TO BIOCHEMISTRY

Many biological molecules contain iron at their active centres.
Since the Mössbauer effect provides a powerful probe of the
chemical state and the environment of iron atoms, it is clear
that it can be a useful tool to be applied to the study of
proteins.

Proteins are large molecules, with molecular weights typically
between 10^4 and 10^6. They contain a large number of amino-
acids joined together by polypeptide bonds and coiled around
to form a helix. The structure of several of them has been
determined by X-ray diffraction. This is a long and difficult
process but it requires proteins which can be crystallized and
in which heavy atoms can be substituted in order to determine
the phase of the diffracted X-rays. Measurements of Mössbauer
spectra do not require single crystals and may be used a) to
study the chemical state and bonding and b) to obtain qualitative data on the local structure and symmetry in the neighbour-
hood of the iron atoms.

The measurements are usually made on frozen aqueous solutions
of the proteins, as this is usually the simplest solid form
which is stable and easily obtainable. Some measurements
have been made on concentrated proteins which have been
precipitated and separated in solid form using a high speed
centrifuge.

The molecules are generally enriched in Fe^{57}, since natural
iron contains only 2% of the Mössbauer isotope, although a few
measurements have been done using naturally occurring iron.
The Fe^{57} may be incorporated in the molecule in two ways;
either by growing the organism from which the protein is
extracted on the separated isotope Fe^{57}, or by incorporating it
by chemical exchange. The growing method requires more Fe^{57},
but it is more reliable and more generally applicable than
exchange, which must be tested carefully to ensure that the
protein is not modified by this process.

The main groups of biological molecules which contain iron at their active centres are shown in Figure 10.1.

The haem proteins are the best understood of these molecules, and the first systematic study of biological molecules using the Mössbauer effect was done on haemoglobin and its derivatives by Lang and Marshall (1966). Since then a great deal of work has been done on the iron-sulphur proteins and on iron storage proteins. In principal useful biological information could be obtained with other Mössbauer isotopes (e.g. I^{127}, I^{129}) but so far no data have been reported on proteins containing them.

Each of the three main features of the Mössbauer spectrum

    a)   The isomer (chemical) shift.
    b)   The quadrupole splitting.
    c)   The magnetic hyperfine splitting.

gives different and independent information about the electrons of the iron, and is sensitive (especially the magnetic hyperfine structure) to small details of the electron wave functions.

Because of the effects of covalency the shifts and splittings observed in the Mössbauer spectrum may be different in biological molecules compared with inorganic complexes. Although the isomer shift depends upon the oxidation state and degree of covalency of the iron, it is not always possible to use it unambiguously to measure the oxidation state in proteins. There is no general theory of the isomer shift, so an empirical calibration of it is necessary.

In inorganic salts the shift decreases as the degree of covalency of the ligands increases, i.e. in the order $-H_2O$, $-Cl^-$, $-O^{2-}$, $-S^{2-}$, $-CN^-$ etc. Also it is systematically less by about 0.2 mm/sec for tetrahedral co-ordination compared with octahedral co-ordination to the same ligands. So the chemical state cannot be inferred from the value of the isomer shift alone. A possible exception might be high spin Fe^{2+} which normally has a large positive shift value, but in biological compounds with tetrahedral sulphur co-ordination (e.g. reduced rubredoxin) it can be as low as 0.6 mm/sec which overlaps with the values found for Fe^{3+} in other compounds.

The quadrupole splitting may be used to deduce the nature of the distortion of the neighbouring atoms from cubic symmetry,

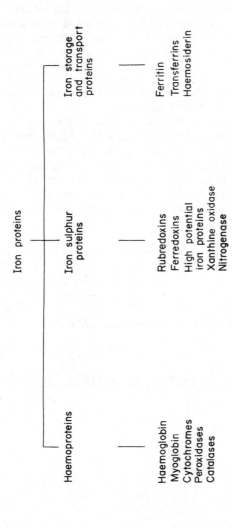

Fig. 10.1 Diagram showing the main groups of biological molecules which contain iron.

i.e. it provides qualitative data on the local molecular structure close to the iron atom. It may also provide confirmation of the chemical state as determined from chemical shift data.

Magnetic hyperfine splitting is sensitive to the state of the iron atoms. It may be observed at low temperatures even in zero applied field for Fe^{3+} atoms where the electron spin-lattice relaxation time (T_1) is long. The electron spin-spin relaxation time (T_2) is generally long in proteins since they are magnetically dilute. For Fe^{2+} atoms the electron spin-lattice relaxation time is short and it is necessary to apply a strong magnetic field at low temperatures in order to observe magnetic splittings. For slowly relaxing paramagnetic ions the application of an external magnetic field usually produces a simpler Mössbauer spectrum, which is easier to interpret than the zero-field spectrum. The magnetic hyperfine interaction is a tensor quantity, and when its anisotropy can be measured it enables (as does the sign of the quadrupole splitting) the orbital wave function of the iron d-electrons to be determined.

High-spin Fe^{3+} has an isotropic magnetic hyperfine interaction, correspondening to an effective field at the nuclei of the order of 500 kG. Low-spin Fe^{3+} and high-spin Fe^{2+} have anisotropic hyperfine interaction and in general their spectra show broader lines than high-spin Fe^{3+}. Low-spin Fe^{2+} has S=0 and hence shows no magnetic hyperfine effects.

REFERENCES

Lang, G. and Marshall, W. (1966). Proc. Phys. Soc. 87 3.

CHAPTER XI

THE USE OF MÖSSBAUER SPECTROSCOPY IN ARCHAEOLOGY AND THE FINE ARTS

Introduction

Archaeology is concerned with the study of the material remains of man's past history and activities. These remains may be fossils, tools, utensils, ornaments or monuments. From his investigations the archaeologist hopes to understand something of the development and spread of the different cultures and technologies. In addition to the written records which exist for about the last 5000 years, techniques for examining ancient objects are required, so that these may be fully characterised and their significance assessed. In the past 20-30 years several such techniques have been developed in the natural sciences, which fall into three main groups.

Firstly there are techniques for determining chemical and physical composition. The results of applying such techniques to many objects of a given type can help to reveal the sites of manufacture and also the trade routes by which they were distributed. An example of such a technique is X ray fluorescence where X rays characteristic of a given element are excited by bombarding the object with either X rays or gamma rays of higher energy than the characteristic X rays.

In another set of measurements, the age of a given object may be determined, using for example the decay of radioactive carbon-14 in objects containing carbon.

A combination of these two measurements has been used to distinguish between 'genuine' ancient artefacts and 'modern' reproductions or fakes.

The third main use of scientific techniques is in surveying likely sites to locate buried objects and to tell the archaeologist which are the most profitable places to dig. An example of this is magnetic surveying in which one detects the small local anomalies in the intensity of the earth's magnetic field associated with buried objects or products of past human habitation.

There have been several publications in which Mössbauer spectroscopy has been used to analyse the composition of archaeological objects containing iron. Although Mössbauer resonances have been observed in about 40 nuclides, measurements in this field have used mostly the most favourable one, iron-57. Several techniques, X ray fluorescence is an example, yield an elemental analysis, but the Mössbauer effect may be used to identify the chemical forms of any iron present and their proportions. The technique is also successful when applied to finely divided iron compounds which give no diffraction lines in an X ray diffraction pattern.

Experimental techniques

The basis for the identification of iron compounds present in a particular sample is the measurement of the hyperfine interactions - the isomer shift, quadrupole coupling and magnetic hyperfine interaction, and their comparison with those already measured for likely compounds. Thus a data bank is needed containing the spectra and hyperfine parameters of many compounds. For studies of clay minerals used in the manufacture of pottery identification is difficult because the minerals exist as solid solutions over a wide range of compositions and can contain various impurities or defects. Thus the number of reference spectra required is very large and usually a conclusive identification is impossible although the type of compound may be distinguished from the overall shape of the spectrum. The search may be narrowed down if there are also available the results of an elemental analysis.

The Mössbauer isotopes which easily yield spectra at room temperature are iron-57, tin-119, and europium-151, of which the first two are clearly significant in archaeological materials. Iron is present in many pottery samples and cutting implements made of obsidian or flint while both iron and tin may be present in tools and coins. If the measuring technique is to be wholly non-destructive then one must either have an article such as a potsherd which is suitable for a Mössbauer transmission measurement (several mms thick) or else measure a backscattered Mössbauer spectrum characteristic of the outer

surface layers of the sample, such as the glaze on a pot
surface. In the study of pottery, where the iron concentration can be about 5%, it is sometimes possible to extract
about 100 mg. of powder which is sufficient to obtain a
reasonable transmission spectrum. This powder, or in some
cases the whole sherd, may be cooled to 77K or 4.2K in order
to increase the recoilless fraction f or to use the magnetic
properties as an aid to identification. From this information
an idea of the particle sizes of finely divided material may
sometimes be obtained.

For samples containing a low concentration of iron and perhaps
significant amounts of heavy elements which produce a relatively high background in the Mössbauer transmission spectrum,
it is necessary to use a radiation detector with high resolution and high signal/background at 14 keV. A proportional
counter filled with argon/methane is normally used with the
strongest source of ^{57}Co available (typically 50-100 mCi).

The backscattering method relies in the case of ^{57}Fe on the
detection of the decay of the 14 keV level in the sample
(scatterer) via emission of either 14 keV gamma rays, 6.3 keV
K conversion X rays or the conversion electrons with energies
up to about 7 keV. These radiations will have different
penetrating powers in the sample used, of about 0.1 μm, 0.1 mm,
and 1 mm for electrons, X rays and gamma rays, in a typical
pottery sample. These ranges then define the depth below the
surface from which the radiation can escape. Thus the spectra
will be characteristic of the iron compounds within different
depths according to the type of scattered radiation detected.
X ray and gamma ray scattered spectra are usually most useful,
because the spectra obtained using electrons will probably
display the effects of weathering. Proportional counters
to detect either the scattered gamma or X rays have been
constructed, having the main design features set by the need to
detect as much of the required radiation as possible and as
little of the background radiations resulting from photoelectric
absorption. Although the counting rates obtained in scattering
are in general an order of magnitude less than in transmission,
in principle one can observe larger resonant effects, limited
not by the recoilless fraction as in transmission, but by the
relative background and signal cross-sections.

Studies of pottery

In the study of pottery the archaeologist is interested in
knowing the places of manufacture or provenance, the techniques
used in making and decorating the pots and in the ways in which
the various styles developed and spread through trade.

Mössbauer analysis of pottery samples (Gangas et al 1973,
Kostikas et al. 1974) has shown that unfired material contains
iron as finely divided oxides and also as ferric and ferrous
ions incorporated in the clay mineral (typically silicate)
structure. On firing there is a transformation of the oxides
to different forms e.g. $\beta FeOOH$ to αFe_2O_3, some aggregation of
the particles leading to changes in the superparamagnetic
behaviour and eventually at about $800^\circ C$ a transformation of
the clay mineral structure.

The occurrence of finely divided particles ($\simeq 100 \text{Å}$) containing
iron leads to superparamagnetism. All the spins on the iron
atoms within a particle point in the same direction, but the
influence of thermal fluctuations causes this direction to vary
between the different easy directions of magnetisation with a
frequency depending upon the particle size, anisotropy energy
and temperature. If this frequency is greater than the Larmor
precession frequency of the ^{57}Fe nucleus (about 4×10^7 sec^{-1}
in 500 kOe) then the hyperfine splitting will collapse to give
a single line, or a doublet in the presence of a quadrupole
interaction. In the opposite case of slow relaxation of the
iron spins, the full hyperfine splitting is observed. This is
illustrated in fig. 11.1 where the relaxation rate is reduced by
reducing the temperature.

A comparison of the Mössbauer spectra of well authenticated
samples of Mycenean and Minoan pottery has been started
(Kostikas et al, 1974). Here the provenance is usually well
known and there exist different but essentially homogeneous
styles. The spectra (fig. 11.2) are roughly similar although the
amount of magnetic component (outer six lines) varies as does
the proportion of structural ferrous ions, as indicated by the
line at about 2 mms^{-1}. It has not so far been possible to
correlate these differences with changes in the amount of
decoration, colour or texture but such correlations may emerge
when many samples have been examined. On the other hand,
computer fitting of the spectra has revealed an interesting
difference in the quadrupole splitting of the central ferric

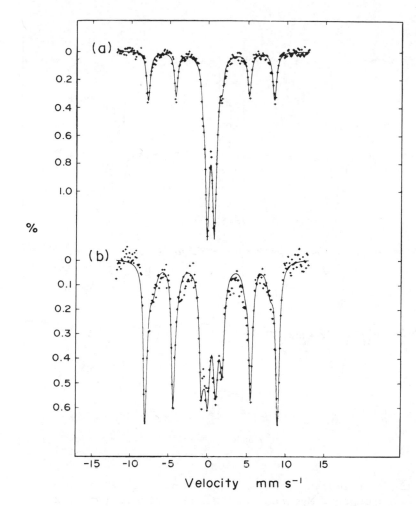

Fig. 11.1 Mössbauer absorption spectra of the body of a Greek 'Etruscan' potsherd at room temperature (300 K) (a) and at 4.2 K (b). The full lines are least squares fits to the data. In all figures, the observations were obtained using a [57]Co source in a rhodium matrix.

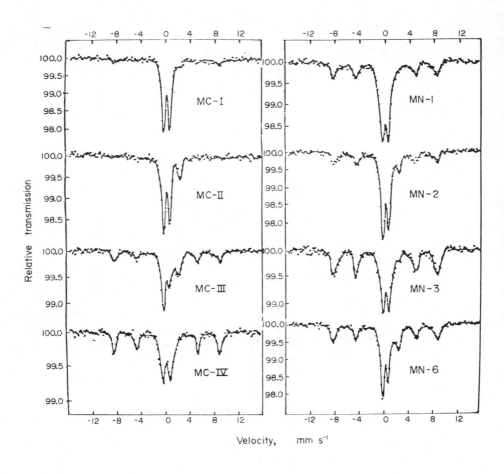

Fig. 11.2 Representative Mössbauer spectra at 77K of two
groups of Mycenean and Minoan pottery (Kostikas
et al 1974)

doublet between the Mycenean and Minoan samples. (fig. 11.3)

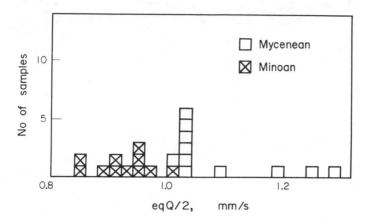

Fig. 11.3   Values of quadrupole splitting for samples of
Minoan X and Mycenean pottery. (Kostikas
et al 1974)

Another application of the Mössbauer effect has been to
identify the black pigment responsible for the surface gloss
on Greek Black vases. (Longworth and Warren, 1975). In this
type of pottery the red body of the pot is covered with a
metallic-like black gloss which was produced from the same
clay as the body with apparently no additional materials. It
was suggested that the colours were obtained in a single
firing with the kiln atmosphere being oxidising, then reducing
and finally oxidising again. The red colour could be due to
ferric oxide $\alpha-Fe_2O_3$, (haematite) while the black colour was
thought to be due to the reduced forms, either magnetite Fe_3O_4
or wüstite FeO, which survived the final oxidation through
being finely divided and partially sintered so as to be
effectively sealed from the oxygen in the air. Because of the
small size of these particles the oxides could not be identified by X-ray diffraction. Fig. 11.4 shows the Mössbauer
spectrum for the black gloss on a sherd of 6th Century Greek
Black pottery at several temperatures. The room temperature
spectrum is similar to that for the red core of the sherd (fig.
11.1), consistent with the supposition that only one clay is
involved. However the six line magnetic spectrum for the core
corresponds to that of ferric oxide $\alpha-Fe_2O_3$, while the 12 line

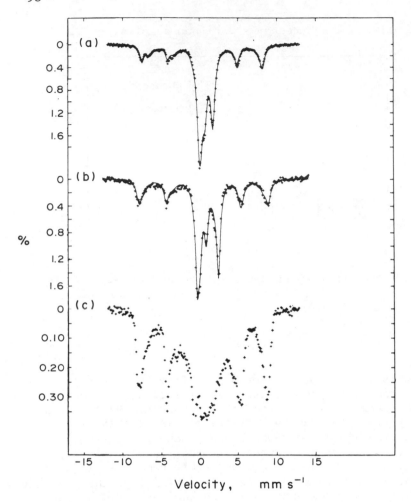

Fig.11.4 Mössbauer absorption spectra of material taken from
the surface of a Greek 'Etruscan' potsherd (sample
2) at room temperature (a), 77 K (b) and 4.2K (c).

pattern in the spectrum for the gloss corresponds reasonably
with that expected for magnetite Fe_3O_4. These assignments
are clearly in agreement with the observed colours of the
gloss and core. The absorption spectrum for the gloss was
obtained by removing about 20 mg. of material and was found
to be similar to a Mössbauer X-ray scattered spectrum for the
gloss. Although this latter technique is non-destructive,

counting rates are typically an order of magnitude down on
those for absorption spectra while the percentage effect in
the spectra are comparable.

Study of soils at archaeological sites

Buried pits and ditches at archaeological sites can often be
located by the increase in magnetic susceptibility of the soil
filling with respect to that of the sub-soil, (Tite and
Mullins 1971), which gives rise to an anomaly in the local
magnetic field intensity. This increase is thought to be due
to the conversion of the weakly magnetic oxide, haematite
α-Fe$_2$O$_3$ to the strongly ferrimagnetic oxide, maghaemite
γ-Fe$_2$O$_3$, by reduction and partial re-oxidation. The mechanism
involved may be either a natural one using the decay of organic
matter under wet conditions to produce reduction followed by
oxidation under dry conditions, or man-made, in which case the
reduction process is due to the burning of organic matter at
the site during human occupation. These processes have been
studied by comparing the Mössbauer spectra for samples of
sedimentary marl (OXN 14 ASR and NORD 8 ASR) with that for
sand (DR 19 ASR) from an Iron Age Roman Settlement (fig. 11.5)
(Longworth and Tite, 1977) Also shown are the spectra for the
iron oxides, geothite, haematite, magnetite and maghaemite.
In these measurements at 4.2K an external magnetic field of
30kOe was applied to the samples in order to distinguish
antiferromagnetic haematite from ferrimagnetic maghaemite.
The application of a field to an antiferromagnet such as
haematite or goethite merely broadens the lines without
changing the overall splitting. For a ferromagnet or ferri-
magnet such as maghaemite or magnetite the intensities of
lines 2 and 5 in the spectrum are greatly reduced when a field
is applied in the gamma ray propagation direction. In this
way samples NORD8 ASR and OXN 14 ASR were shown to contain
haematite while sample DR 19 ASR contained maghaemite. Sample
NORD8 HTD consists of a sample of NORD8 ASR which had been
heated at 550°C in a nitrogen atmosphere followed by air, in
the laboratory in order to simulate the burning mechanism.
Again the spectrum is very similar to that for maghaemite as
opposed to the original haematite in NORD 8 ASR.

Studies of artists' pigments

Many artists' pigments are known to contain iron, usually in
the form of oxides which are mainly responsible for the

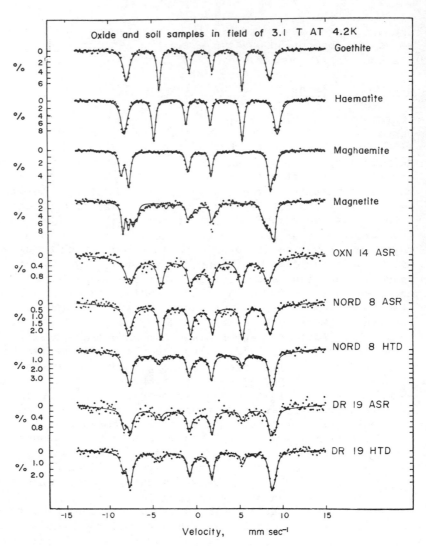

Fig.11.5 57Mossbauer spectra for soil and standard oxides at
4.2K in a magnetic field of 3.1T along the gamma ray
direction.

colours involved. Mössbauer measurements on pigments may
sometimes be used therefore to distinguish between natural and
synthetic pigments on the basis of particle size, and perhaps
aid in settling authenticity problems. Many pigments have
been analysed in this way using transmission measurements
(Keisch, 1974) or backscattered X-rays (Longworth and Dale 1973).

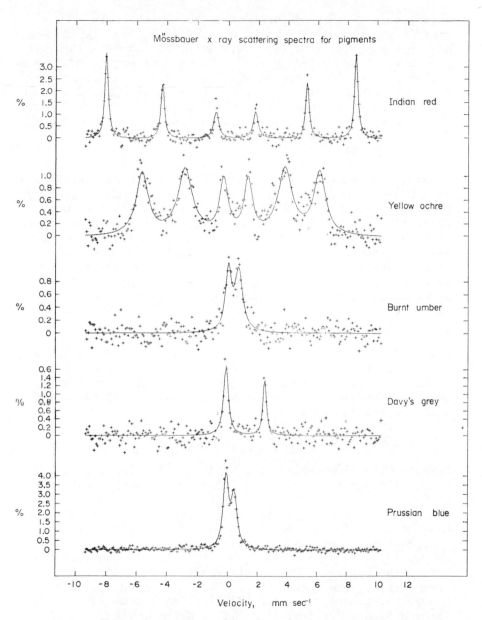

Fig. 11.6 Mössbauer backscatter spectra of artists' pigments

This latter type of measurement is of course truly non-destructive and ideally could be carried out on a given area of a painting in situ. Fig. 11.6 shows the spectra for several pigments, of which Indian Red is largely haematite, while the yellow iron oxides, including Yellow Ochre, contain the hydrated form α-FeOOH. Burnt Umber is an example of a brown pigment containing haematite this time in finely divided form. Davy's Grey contains iron in a silicate structure while Prussian Blue is a synthetic pigment consisting of a ferrocyanide complex.

Although there have been relatively few Mössbauer measurements on archaeological objects, they do serve to demonstrate the potential uses of the technique in this field. Much work needs to be done to produce Mössbauer spectra of many examples of each type of object before it will be possible to obtain useful information, such as the distinction of different manufacturing sites. It is likely that Mössbauer measurements using tin-119 to examine bronzes will prove useful although the size of the hyperfine parameters with respect to the linewidth of the resonance is not as favourable as for iron-57.

The use of Mössbauer spectroscopy by itself in archaeology is clearly limited, but when combined with other complementary techniques, such as X-ray fluorescence, neutron activation or thermoluminescence, it can prove a powerful tool for identification and analysis.

REFERENCES

Gangas, N.H., Simopoulos, A., Kostikas, A., Yassoglou, N.J. and Filippakis, S. (1973). Clays and Clay Minerals, 21, 151.

Keisch, B., (1974). Journal de Physique, Coll. C1. 35, 151.

Kostikas, A., Simopoulos, A and Gangas, N.H. (1974). Journal de Physique, Coll. C1. 35, C1-107.

Longworth, G. and Dale, B.W. A.E.R.E. Harwell Nuclear Physics Progress Report 1973-74, AERE-PR/NP21.

Longworth, G. and Warren, S.E. (1975). Nature, 255, 625.

Longworth, G. and Tite, M.S. (1977). Archaeometry, 19, 3.

Tite, M.S. and Mullins, C. (1971). Archaeometry, 13, 209.

CHAPTER XII

APPLICATIONS TO SURFACE SCIENCE

Surface Science is the study of the first few atomic layers of
a solid and is of interest as a fundamental topic in its own
right and for its applications, for example to catalysis.
Although Mössbauer spectroscopy is not specifically a surface
technique, as is for example LEED (Low Energy Electron
Diffraction) and other techniques using low energy electrons
which have only a very short absorption length in matter, it
is possible with ingenuity to design experiments to study
surfaces. Measurements are commonly carried out on surfaces
prepared in one of three ways:

(i)   Thin films containing a Mössbauer absorbing isotope.
       These may be deposited on a substrate which may be a
       metal or an insulating material. It may be necessary to
       mount several films on top of each other (provided the
       substrate is also thin) in order to achieve a reasonable
       absorption.

(ii)   Fine particles. These have most of their atoms at or
       near the surface and so if they contain a Mössbauer
       isotope can be used to study the surface. It is also
       sometimes possible to enrich the surface layer in the
       Mössbauer isotope. In measurements on fine particles it
       is important to distinguish effects due to the surface
       from those due to the small size of particles
       themselves, e.g. superparamagnetic effects. (see p.94)

(iii) Sources. In this case, Mössbauer emitting nuclei are
       deposited on the surface to be studied.

When it is possible to prepare the same surface in different
ways, data from each kind of measurement are in good agreement
with each other.

In true surface techniques the specimen is studied in ultra high vacuum uhv ($< 10^{-10}$ torr) to avoid adsorption of, and reaction with, residual gas atoms. These conditions can also be achieved by the thin film technique, particularly if the film is covered for protection with another film of a material not containing the Mössbauer isotope. Sandwiches of alternate layers of, for example, copper and iron have been prepared. In such a case, of course, it is the solid-solid interface which is studied rather than solid vacuum. Fine particle and source specimens can also sometimes be coated with protective layers.

Fine particles of dimension 10 nm or less are usually prepared by evaporation of, or precipitation from, aqueous solution, and uhv conditions are unlikely to be realised in their preparation. Although these samples do not allow a true surface measurement, data on them may be extremely valuable for the study of catalysis. Catalytic processes, which were the original motivation for the development of surface science, take place at the surface of the catalyst and most solid catalysts are in the form of fine particles. Unlike a true surface technique, Mössbauer spectroscopy can give information on fine particles in the same environment as is used in a catalytic reaction.

An atom close to the surface of a solid would be expected to have different properties from an atom in the bulk of the sample owing to the imbalance of the forces acting on it - a similar imbalance in liquids gives rise to surface tension. There is a certain arbitrariness about what is meant by close to the surface, but in practice it is believed that the surface is the first few atomic layers, say up to five, or to a depth of about 1 nm. Very thin films or fine particles (< 1 nm) would therefore be expected to show Mössbauer spectra characteristic of the surface only. Larger systems (1-10 nm) would show contributions from both surface and bulk. In systems larger than 20 nm the surface fraction would be too small to detect. Measurements on such samples, using the "back-scatter" technique are discussed in the next chapter.

Atoms near the surface therefore have a slightly different electron distribution and a lower symmetry compared to those in the bulk. Consequently, their Mössbauer spectra could be expected to show a different chemical shift, and they might also show a quadrupole splitting even in a crystal of overall cubic symmetry. In magnetic solids the surface atoms might be expected to have different magnetic moments, and according to one theory the magnetic moments in the surface layer would be zero, the so called "dead" layer.

Measurements on thin films of α-Fe sandwiched between layers of silver and copper (Duncan et al. 1978, Keune et al. 1979) showed that the isomer shift was slightly greater than bulk iron, i.e. the electron density at the nuclei decreased near the surface. This is illustrated in Fig. 12.1 taken from Owens et al. (1979), for films of around 20 layers of iron (about 6 nm) sandwiched between films of silver (about 200 nm). In these experiments ^{57}Fe and ^{56}Fe were deposited so that the ^{57}Fe occurred at different depths in different specimens.

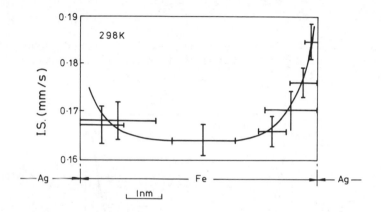

Fig. 12.1. Isomer shift variation at 298K as a function of depth in films of iron in silver. (Owens et al. 1979). Results at other temperatures are similar.

The hyperfine spectra at room temperature of α-iron sandwiched between several substrates (copper, silver, MgF_2 MgO) have been measured. It is found that atoms in the surface have a smaller field than atoms in the bulk, i.e. the magnetic moment a the surface is smaller than in the bulk. This is shown in Fig. 12.2 for films in silver. Such behaviour is to be expected on basis of the molecular field model – the molecular field and hence the atomic moment (and hyperfine field) decrease as the surface is approached.

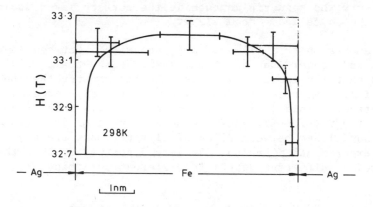

Fig. 12.2. Magnetic hyperfine fields at 298K as a function of depth in films of iron on silver (Owens et al. 1979).

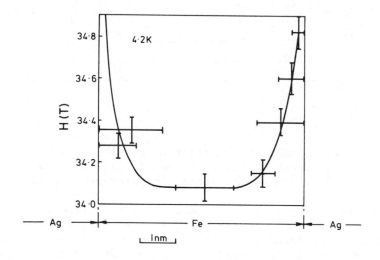

Fig. 12.3. Magnetic hyperfine fields at 4.2K as a function of depth in films of iron on silver (Owens et al. 1979).

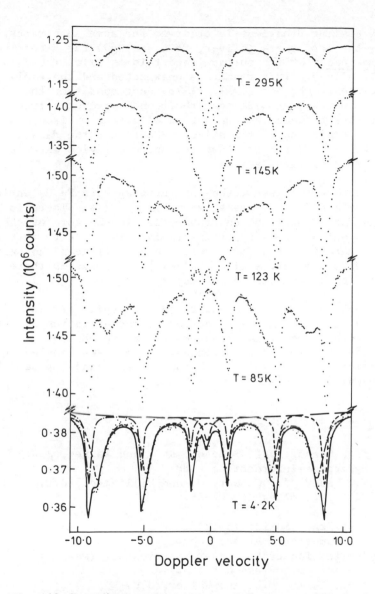

Fig. 12.4. Mössbauer spectra of 7 nm particles of α-Fe$_2$O$_3$ enriched in [57]Fe at the surface at several temperatures (van der Kraan 1973).

At 4.2K similar behaviour is observed for some substrates, e.g. copper, but for others (silver, MgF_2 and MgO) the hyperfine field <u>increases</u> at the surface, and this is illustrated in Fig. 12.3 for silver. This result is unexpected and the explanation not yet clear. Duncan et al. (1978) suggested that the observed increase at 4.2K near the boundary results from interactions between the conduction electrons of iron and the outer electrons of the silver. Note that the data in Figs. 12.1 - 12.3 confirm that the surface region extends down to four or five layers.

Data on fine particles of the antiferromagnet α-Fe_2O_3 enriched at the surface in ^{57}Fe are shown in Fig. 12.4. There is evidence for a contribution from nuclei in the core of the particles which have a hyperfine field very close to that of the bulk material, and from nuclei in a shell near the surface which have a somewhat smaller value. In no case is there any evidence for a magnetic "dead" layer.

A large volume of data now exists on the topics already mentioned and also on the direction of magnetization and on the motion of the atoms (via measurements of the Mössbauer recoilless fraction f) at surfaces. For reviews, including applications to catalysis see Mørup, Dumesic and Topsøe (1980) and Topsøe, Dumesic and Mørup (1980).

REFERENCES

Duncan, S., Owens, A.H., Simpson, R.J. and Walker, J.C.
(1978) Hyperfine Interactions 4 88
Keune, W., Lauer, J., Gonser, U. and Williamson, D.L.
(1979) J. Phys. Colloq. 40 C2-69.
Kraan, A.M. van der,
(1983) Phys.Stat. Solid: 18 (a) 215.
Mørup, S., Dumesic, J.A. and Topsøe, H.
(1980) Application of Mössbauer Spectroscopy (ed.
R.L. Cohen Academic Press N.Y.) p.1.
Owens, A.H., Chien, C.L. and Walker, J.C.
(1979) J. Phys. Colloq. 40 C2-74.
Topsóe, H., Dumesic, J.A. and Mørup, S.
(1980) Application of Mössbauer Spectroscopy (ed.
R.L. Cohen Academic Press N.Y.) p.55.

CHAPTER XIII

APPLICATIONS OF BACKSCATTERING MEASUREMENTS

A Mössbauer scattering spectrum is measured by counting the
number of either gamma rays, X rays or electrons emitted in
the decay of the Mössbauer level in the scatterer nuclei, as a
function of source velocity. Although these decay modes are
in general non-resonant processes, they result from a
Mössbauer absorption and hence can be used to measure a
Mössbauer spectrum. Most of the early scattering work was
concerned with the scattering of Mössbauer gamma rays in
experiments similar to X ray diffraction measurements, in
which the interference effects between the scattering from a
regular array of atoms are involved. More recently the
emphasis has moved to measurements on scattered electrons or X
rays, sometimes known as Conversion Electron Mössbauer
Scattering (CEMS) and Conversion X ray Mössbauer Scattering
(CXMS).

In the case of iron-57, for every 100 Mössbauer gamma ray
absorptions, only about 9 result in the re-emission of the
14.4 kev gamma ray, while 81 and 9 result in the emission of K
(7.3 kev) and L (13.6 kev) conversion electrons. Emission of
these conversion electrons is followed by the emission of K
conversion X rays and L Auger electrons, in the ratio 27:63.
The escape depths for the gamma rays, X rays and K electrons
in iron are about $20\mu m$, $15\mu m$ and 100 nm. The conversion and
Auger electrons may be used therefore to study near surface
effects and a backscattering geometry is frequently used (Fig
13.1) in which the sample is incorporated as part of the back
surface of a proportional counter.

In the integral scattering measurement the electrons leaving
the surface of the scatterer with any energy between close to
zero and 13.6 kev are counted, with the result that there is
little depth discrimination in the measured spectrum. If the
scattered electrons are analysed into energy bands by an
electron spectrometer, and spectra are taken for each band,
then provided the curves for electron energy loss against

thickness are known, it is possible to calculate the spectrum arising from atoms at a given depth. This depth selective technique has the disadvantage that the counting geometry is much poorer than the 2π geometry in the integral technique, and counting times are necessarily longer.

A fuller survey of experiments of both types has been given by Tricker 1981. Here four examples are given of the use of the integral technique to study steels subjected to either corrosion or cold working, and ion implanted alloys.

1.   The hyperfine fields at iron atoms in iron metal and in magnetic iron oxides such as magnetite (Fe_3O_4) are very different (about   30T and 50T), so that they are easily identifiable. This means that the formation of oxide layers on ferrous alloys, for example due to corrosion, is detected. Fig. 13.2 shows electron scattering (A,B,C) and X ray scattering (D) spectra for a sample of steel containing 9% chromium, which was oxidised in CO_2 based atmospheres at 550C (Longworth 1977) The spectrum (A) of the original steel is made up of five components, each of six line magnetic patterns, arising from iron atoms with differing numbers of chromium near neighbours.

Steel samples which showed protective oxidation, were shown to have developed a two layer oxide, with the boundary between the two layers being at the original steel surface. The electron scattered spectra (B,C) showed that the outer layer was an iron spinel (magnetite) while the inner one was an iron-chromium spinel having the same crystal structure. The spectrum for the inner oxide layer was obtained after the appropriate thickness had been removed from the sample surface. An X ray scattered spectrum (D) of the oxidised steel which probes a depth 10-100 times greater than the oxide thickness, shows a similar pattern to that of the original steel. However the smaller number of components indicates that there is now less chromium in the layer contributing to the Mössbauer spectrum.

It is deduced that during oxidation iron and chromium atoms diffuse out to the oxide/gas interface while oxygen diffuses inwards. The observed sequence of surface layers may be explained if the iron atoms diffuse further than the chromium atoms.

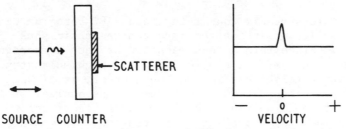

SOURCE COUNTER

Fig 13.1 Schematic diagram of Mössbauer backscattering set-up.

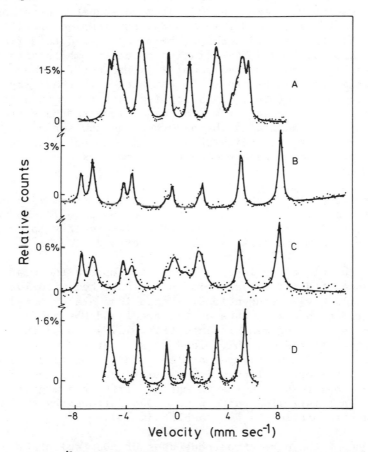

Fig. 13.2 Mössbauer scattered spectra for 9% chromium in steel oxidised in CO_2 A-original steel (CEMS), B,C-outer and inner oxide layers (CEMS) and D-near surface below oxide layers (CXMS)

2.   Stainless steels containing about 18% chromium and 8% nickel usually consist of the face centred cubic phase, austenite. Since this is non-magnetic at room temperature its Mössbauer spectrum is a single line. If such an austenitic steel, with a relatively low carbon content, is either quenched from about 1000C to sub-zero temperatures, or is subjected to cold working, then a martensitic transformation occurs to body centred cubic α'-martensite. This phase is ferromagnetic at room temperature and produces a six line Mössbauer pattern with a hyperfine field H $\simeq$ 30T. This transformation has been studied using Mössbauer scattering, Fig. 13.3 (Bowkett and Harries 1978), for samples of AISI steel subjected to cold working, by being rolled to a final thickness of 0.84 mm. The relative amount of martensite in the more highly cold worked samples is greater for the electron scattered spectra (3,5) compared with the X ray scattered spectra (4,6). This implies that the martensite concentration decreases from its value close to the surface. If a sample is chemically thinned then the new surface concentration of martensite remains high, emphasizing the influence of the surface energy on the martensitic transformation.

A further example of the use of Mössbauer scattering to study phase transformations in steels is given on page 72, in which the effect of wear on the near surface layers of samples of case hardened martensitic steel is described.

3.   A technique developed recently to improve the wear resistance of steel surfaces is to implant them with low energy ($\simeq 10^2$ kev) nitrogen ions. These ions are expected to remain close to the surface of the steel and Mössbauer electron scattering measurements (Fig. 13.4) showed that iron nitrides were produced which lead to an improved wear resistance. (Longworth and Hartley 1978). Spectrum (1) was taken for the original pure iron while increasing doses of nitrogen ions result in the formation first of a magnetic nitride Fe_4N (six line pattern in (2)), followed by the non-magnetic nitride Fe_2N (doublet in (3)).

4.   One of the other important uses of ion implantation is to produce alloys that are not attainable by more conventional techniques. This is possible because ion implantation is a non-equilibrium process whereby large amounts of impurity

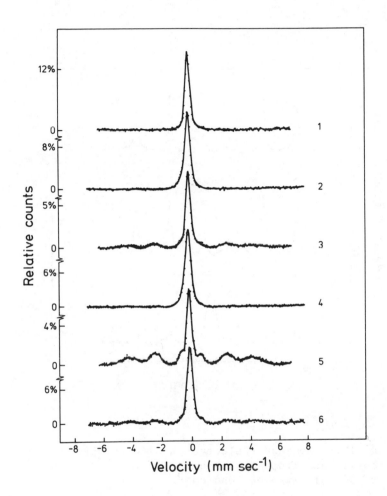

Fig. 13.3 Mössbauer scattered spectra of AISI 321 steel after 5.7% cold work (1,2), 19.5% cold work (3,4) and 50% cold work (5,6) with odd numbers denoting electron scattering and even numbers, X ray scattering.

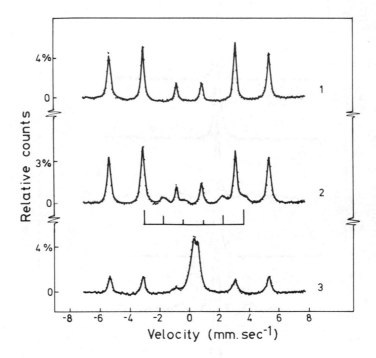

Fig. 13.4 Mössbauer electron scattered spectra for
iron (1) and iron implanted with 2 x 10^{17}(2) and
6 x 10^{17}(3) nitrogen ions cm^{-2}

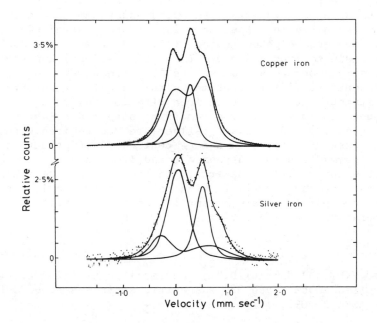

Fig. 13.5 Mössbauer electron scattered spectra
for copper and silver foils implanted with
2×10^{17} iron-57 atoms cm^{-2}

atoms may be introduced into a surface irrespective of the usual equilibrium solid solubilities. Fig. 13.5 shows two examples, of copper and silver respectively, implanted with the Mössbauer atom iron-57, in which the normal solid solubilities have been exceeded. (Longworth 1980). In fact iron is essentially insoluble in silver so that this specimen represents a new alloy. In the spectra (Fig. 13.5) the singlets at about 0.2 mm.s^{-1} (copper) and 0.5mm.s^{-1}, are due to iron dissolved in copper or silver (face centred cubic) with 12 nearest neighbour copper (silver) atoms. The doublet is due again to iron dissolved in copper (silver) but having one or more iron atoms amongst its 12 nearest neighbour atoms. The presence of differing atoms in this shell destroys the cubic symmetry at the central iron site leading to a quadrupole interaction. The remaining singlets at about -0.1 mm.s^{-1} (copper) and 0 mm.s^{-1} (silver are due to iron with 12 iron nearest neighbours in a face centred cubic structure, that is gamma-iron (or austenite), and iron in body centred structure the normal alpha-iron. In the latter case it is supposed that the normal six line pattern for iron is not seen because the particles are so small that they exhibit superparamagnetism (page 94).

REFERENCES

Bowkett M.W. and Harries D.R., "Martensitic transformations in cold rolled EN58B (Type 32) austenitic stainless steel", UKAEA Report AERE R-9093, 1978.

Longworth G., Non Destructive Testing, NDT International, October 1977, p.241.

Longworth G., in "Treatise on Materials Science and Technology", ed. Herman, H., Academic Press: 1980, p.107.

Longworth G. and Hartley N.E.W., Thin Solid Films, 48, 95, 1978.

Tricker M.J. in Mössbauer Spectroscopy and its Chemical Applictions", ed. Stevens J.G. and Shenoy G.K., Advances in Chemistry Series 194, American Chemical Society 1981, p.63.

INDEX